NON-NEUTRAL PLASMA PHYSICS VII

Proceedings in the Series of Workshops on Non-Neutral Plasma Physics

	Year	Publisher	ISBN
VII	2008	AIP Conference Proceedings 1114	978-0-7354-0651-3
VI	2006	AIP Conference Proceedings 862	978-0-7354-0360-4
V	2003	AIP Conference Proceedings 692	0-7354-0165-9
IV	2001	AIP Conference Proceedings 606	0-7354-0050-4
III	1999	AIP Conference Proceedings 498	1-56396-913-0
II	1994	AIP Conference Proceedings 331	1-56396-441-4
I	1988	AIP Conference Proceedings 175	0-88318-375-7

To learn more about these titles, or the AIP Conference Proceedings Series, please visit the webpage http://proceedings.aip.org

NON-NEUTRAL PLASMA PHYSICS VII

9th International Workshop on
Non-Neutral Plasmas

New York, New York   16 – 20 June 2008

EDITORS
James R. Danielson
*University of California
San Diego, California*

Thomas S. Pedersen
*Columbia University
New York, New York*

SPONSORING ORGANIZATIONS
National Science Foundation
Columbia University

Melville, New York, 2009
AIP CONFERENCE PROCEEDINGS ■ 1114

Editors

James R. Danielson
University of California, San Diego
9500 Gilman Dr. 0350
La Jolla, CA 92093-0350

E-mail: jdan@physics.ucsd.edu

Thomas S. Pedersen
Columbia University
500 W. 120th Street
New York, N.Y. 10027

E-mail: tsp22@columbia.edu

Authorization to photocopy items for internal or personal use, beyond the free copying permitted under the 1978 U.S. Copyright Law (see statement below), is granted by the American Institute of Physics for users registered with the Copyright Clearance Center (CCC) Transactional Reporting Service, provided that the base fee of $25.00 per copy is paid directly to CCC, 222 Rosewood Drive, Danvers, MA 01923, USA. For those organizations that have been granted a photocopy license by CCC, a separate system of payment has been arranged. The fee code for users of the Transactional Reporting Services is: 978-0-7354-0651-3/09/$25.00

© 2009 American Institute of Physics

Permission is granted to quote from the AIP Conference Proceedings with the customary acknowledgment of the source. Republication of an article or portions thereof (e.g., extensive excerpts, figures, tables, etc.) in original form or in translation, as well as other types of reuse (e.g., in course packs) require formal permission from AIP and may be subject to fees. As a courtesy, the author of the original proceedings article should be informed of any request for republication/reuse. Permission may be obtained online using Rightslink. Locate the article online at http://proceedings.aip.org, then simply click on the Rightslink icon/"Permission for Reuse" link found in the article abstract. You may also address requests to: AIP Office of Rights and Permissions, Suite 1NO1, 2 Huntington Quadrangle, Melville, NY 11747-4502, USA; Fax: 516-576-2450; Tel.: 516-576-2268; E-mail: rights@aip.org.

L.C. Catalog Card No. 2009923703

ISBN 978-0-7354-0651-3
ISSN 0094-243X

Printed in the United States of America

CONTENTS

Preface..ix
Organizing Committee xi
Group Photo...xiii

SECTION I
ATOMIC PHYSICS

Sheet Fluorescence and Annular Analysis of Ultracold Neutral Plasmas..3
    J. Castro, H. Gao, and T. C. Killian
Using Charged Particle Imaging to Study Ultracold Plasma Expansion..11
    X. L. Zhang, R. S. Fletcher, and S. L. Rolston
Observation of String Ion Cloud in a Linear RF Trap...............19
    M. Aramaki. S. Kameyama, Y. Sakawa, T. Shoji, and A. Kono
Barium Ions for Quantum Computation..............................25
    M. R. Dietrich, A. Avril, R. Bowler, N. Kurz, J. S. Salacka, G. Shu, and B. B. Blinov
Quantum Optical Heating in Sonoluminescence Experiments..........31
    A. Kurcz, A. Capolupo, and A. Beige

SECTION II
TOROIDAL TRAPS

Achieving Long Confinement in a Toroidal Electron Plasma.........39
    J. P. Marler, J. Smoniewski, B. Ha, and M. R. Stoneking
Recent Progress on Toroidal Non-neutral Plasmas Confined on Heliotron Magnetic Surfaces.......................................47
    H. Himura, K. Nakamura, D. Sugimoto, A. Sanpei, S. Masamune, M. Isobe, and F. Sano
Confinement of Pure Electron Plasmas in the CNT Stellerator......55
    T. S. Pedersen, J. W. Berkery, A. H. Boozer, Q. R. Marksteiner, P. W. Brenner, M. Hahn, B. Durand de Gevigney, and X. S. Martin
Studies of a Parallel Force Balance Breaking Instability in a Stellerator 63
    Q. R. Marksteiner, T. S. Pedersen, J. W. Berkery, M. S. Hahn, J. M. Mendez, B. Durand de Gevigney, P. Ennever, D. Boyle, M. Shullman, and H. Himura
Numerical Studies of Transport in the Columbia Non-neutral Torus...........69
    B. Durand de Gevigney, T. S. Pedersen, and A. H. Boozer
Studies of Enhanced Confinement in the Columbia Non-neutral Torus.........75
    P. W. Brenner, T. S. Pedersen, M. Hahn, J. W. Berkery, R. G. Lefrancois, and Q. R. Marksteiner

Pure Electron Equilibrium and Transport Jumps in the Columbia Non-neutral Torus. 81
    M. Hahn, T. S. Pedersen, J. W. Berkery, Q. R. Marksteiner, P. W. Brenner, and B. Durand de Gevigney

SECTION III
COLLECTIVE MODES AND TRANSPORT

Electron Acoustic Waves in Pure Ion Plasmas 89
    F. Anderegg, C. F. Driscoll, D. H. E. Dubin, and T. M. O'Neil
Excitation of High Order Diocotron Modes in the ELTRAP Device. 96
    G. Bettega, B. Paroli, R. Pozzoli, and M. Romé
Using Variable-frequency Asymmetries to Probe the Magnetic Field Dependence of Radial Transport in a Malmberg-Penning Trap 102
    D. L. Eggleston
Turbulent Cascade in Vortex Dynamics of Magnetized Pure Electron Plasmas. 108
    Y. Kawai and Y. Kiwamoto
Collisional Damping of Plasma Waves on a Pure Electron Plasma Column 114
    M. W. Anderson and T. M. O'Neil
Theory and Simulation of Neoclassical Transport Processes with Local Trapping 121
    D. H. E. Dubin
Effect of a Weakly Tilted Magnetic Field on the Equilibrium of Non-neutral Plasmas in a Malmberg-Penning Trap 130
    M. Romé and I. Kotelnikov
Relatvistic Effects on the Radial Equilibrium of Non-neutral Plasmas 136
    M. Romé, I. Kotelnikov, and R. Pozzoli
Stability of Non-neutral Plasma Cylinder Consisting of Magnetized Cold Electrons and of Small Density Fraction of Ions Born at Rest: Non-local Analysis. 142
    Y. N. Yeliseyev

SECTION IV
APPLICATIONS AND SPECIAL TOPICS

Radial Compression of Antiproton Cloud for Production of Ultraslow Antiproton Beams. 157
    N. Kuroda, Y. Nagata, H. A. Torii, D. Barna, J. Eades, D. Horváth, M. Hori, H. Imao, K. Komaki, A. Mohri, M. Shibata, and Y. Yamazaki
Radial Compression of a Non-neutral Plasma in a Non-uniform Magnetic Field of a Cusp Trap. 163
    H. Saitoh, A. Mohri, Y. Enomoto, Y. Kanai, and Y. Yamazaki
Tailored Particle Beams from Single-Component Plasmas 171
    T. R. Weber, J. R. Danielson, and C. M. Surko

**Investigations on Cooling Mechanisms of Highly Charged Ions at
HITRAP** 179
    G. Maero, F. Herfurth, O. Kester, H.-J. Kluge, S. Koszudowski, W. Quint,
    and S. Schwarz

Investigation of Space-charge Phenomena in Gas-filled Penning Traps 185
    S. Sturm, K. Blaum, M. Breitenfeldt, P. Delahaye, A. Herlert,
    L. Schweikhard, and F. Wenander

**Electrodynamics of Neutron Star Magnetosphere: An Example of
Non-neutral Plasma in Astrophysics** 191
    J. Petri

Next Generation Trap for Positron Storage 199
    J. R. Danielson, T. R. Weber, and C. M. Surko

List of Participants 207
Author Index 211

Preface

The 9th International Workshop on Non-Neutral Plasma Physics was held June 16-20, 2008, in New York City on the campus of Columbia University. Covering a large range of interconnected topics, this workshop continues the tradition of the previously very successful workshops in this series held in Aarhus, Denmark (2006), Santa Fe, NM (2003), San Diego, CA (2001), Princeton, NJ (1999), Boulder, CO (1997), Berkeley, CA (1994), Irvine, CA (1992), and Washington, D.C. (1988). Over fifty scientists from the United States, Europe, and Japan participated, with 24 invited talks, 10 contributed talks, and 20 poster presentations. These proceedings include 28 papers, split into four topical groups including: Atomic, Molecular and Optical Physics; Toroidal Traps; Collective Modes and Transport; and Applications and Special Topics.

Ever since the first workshop, there has been a strong relationship between the non-neutral plasma community and atomic physics. Section I contains a number of papers covering a range of topics from atomic, molecular and optical physics. Topics in this section include experiments with ultracold neutral plasmas, strongly-coupled plasmas, trapped ions for quantum computation, and sonoluminescence.

Over the last several years there has been a resurgence in interest in studying non-neutral toroidal plasmas. In Section II, we have papers covering the latest experiments in three different toroidal traps including a simple torus, a heliotron, and a compact stellarator.

The core of non-neutral plasma research continues to be studies of single-species plasmas in Penning-Malmberg traps. This workshop was no exception. Section III includes a wide variety of experiments and theory concerning basic studies of collective modes and plasma transport.

Finally, Section IV highlights some of the many uses of non-neutral plasma techniques for a variety of applications and special topics. This includes experiments with antimatter plasmas, highly charged ions, and astrophysical non-neutral plasmas.

James R. Danielson, Editor
University of California, San Diego

Thomas Sunn Pedersen, Chair
Columbia University

2008 Workshop on Non-Neutral Plasmas

June 16-20, 2008
Columbia University
New York, New York USA

Organizing Committee

Thomas Pedersen, *Chair*
Columbia University

John Bollinger
National Institute of Standards
and Technology

Michael Drewsen
Aarhus University

Fred Driscoll
University of California,
San Diego

Joel Fajans
University of California,
Berkeley

Matt Stoneking, *Program Chair*
Lawrence University

Gerald Gabrielse
Harvard University

Haruhiko Himura
Kyoto Institute of Technology

Tom Killian
Rice University

Lutz Schweikhard
Greifswald University

Cliff Surko
University of California,
San Diego

SECTION I
ATOMIC PHYSICS

Sheet Fluorescence and Annular Analysis of Ultracold Neutral Plasmas

J. Castro, H. Gao and T. C. Killian

Rice University, Department of Physics and Astronomy, 6100 Main St., Houston, Texas, USA 77005

Abstract. Annular analysis of fluorescence imaging measurements on Ultracold Neutral Plasmas (UNPs) is demonstrated. Spatially-resolved fluorescence imaging of the strontium ions produces a spectrum that is Doppler-broadened due to the thermal ion velocity and shifted due to the ion expansion velocity. The fluorescence excitation beam is spatially narrowed into a sheet, allowing for localized analysis of ion temperatures within a volume of the plasma with small density variation. Annular analysis of fluorescence images permits an enhanced signal-to-noise ratio compared to previous fluorescence measurements done in strontium UNPs. Using this technique and analysis, plasma ion temperatures are measured and shown to display characteristics of plasmas with strong coupling such as disorder induced heating and kinetic energy oscillations.

Keywords: strong coupled plasmas, ultracold plasmas, fluorescence imaging
PACS: 52.27.Gr, 52.70.Kz, 33.50.Dq, 32.80.Fb

INTRODUCTION

Ultracold neutral plasmas (UNPs) [1, 2] provide a platform for studying strongly-coupled systems [3, 4, 5, 6, 7, 8] and expanding plasmas [9, 10, 11, 12]. UNPs are characterized by electron temperatures from 1-1000 K, ion temperatures around 1 K, and densities as high as 10^{10} cm^{-3}.

Strongly-coupled plasmas are characterized by having the Coulomb coupling constant, $\Gamma = e^2/(4\pi\varepsilon_0 a k_B T) > 1$, where T is the temperature of a singly-charged plasma species and $a = (4\pi n/3)^{-1/3}$ is the interparticle distance. Experiments in UNPs show that ions equilibrate with $2 < \Gamma_i < 5$ [13], in agreement with theoretical estimates [14]. Rapid heating processes clamp the electron $\Gamma_e \leq 0.2$ [15, 16, 17, 18].

UNPs are created by photoionization of laser-cooled and trapped strontium atoms [19, 20] in a magneto-optical trap (MOT) (Fig. 1). The neutral atom cloud has a Gaussian density distribution, $n(r) = n_0 \exp(-r^2/2\sigma^2)$, where $\sigma \approx 0.6$ mm and $n_0 \approx 6 \times 10^{10}$ cm^{-3}. The number of atoms is typically 2×10^8 with a temperature of a few mK. To obtain larger samples with lower densities, the trap is turned off and the cloud is allowed to expand.

Photoionization of the atoms is performed by two overlapping $\sim$10 ns laser pulses: one from an amplified laser tuned to the transition used for cooling the atoms, the other from a dye laser tuned just above the ionization continuum (Fig. 2 A). The density profiles of the plasma components, $n_e(r) \approx n_i(r)$, follow the Gaussian shape of the neutral atom cloud, with peak electron and ion densities as high as $n_{0e} \approx n_{0i} \approx 4.5 \times 10^{10}$ cm^{-3}, corresponding to an ionization fraction of 75%.

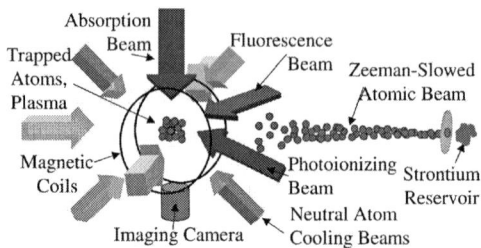

FIGURE 1. Experimental set-up for ultracold neutral plasma experiments. Neutral strontium atoms from a reservoir are Zeeman-slowed and enter the trapping region. The magneto-optical trap (MOT) consists of a pair of anti-Helmholtz magnetic coils and six laser-cooling beams. $^1P_1$ trapped atoms are then ionized by the photoionizing laser. The fluorescence probe beam propagates in a direction that is perpendicular to the imaging axis and CCD camera. The complementary absorption probe beam passes through the plasma and falls on the camera.

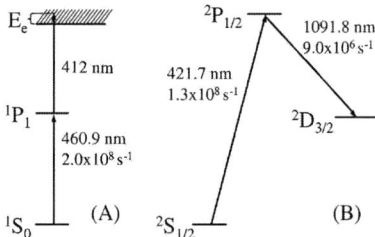

FIGURE 2. Atomic and ionic energy levels involved in strontium experiments, with decay rates. (A) Neutral atoms are laser-cooled and trapped in a magneto-optical trap (MOT) operating on the $^1S_0 - {}^1P_1$ transition at 460.9 nm, as described in [20]. Atoms are later excited to the $^1P_1$ level by a pulsed amplified laser and ionized by photons from a laser at ~ 412 nm. (B) Imaging of the ions is done on the $^2S_{1/2} - {}^2P_{1/2}$ transition at 421.7 nm. $^2P_{1/2}$ ions decay to the $^2D_{3/2}$ state 7% of the time, after which they cease to interact with the imaging beam. The intensity and duration of the 422 nm light is low enough to avoid optical pumping to this metastable $^2D_{3/2}$ state.

PLASMA DYNAMICS

Upon creation of the plasma the electrons have an initial kinetic energy, E_e, approximately equal to the difference between the photon energy and the ionization potential. This is a consequence of the small electron-ion mass ratio. The photoionizing beam is detuned above the ionization threshold, such that E_e is typically set from 1-1000 K. Electrons mix rapidly, and the thermal equilibrium is nearly global. The equilibration time is tens of nanoseconds with a corresponding temperature of $T_e \approx 2E_e/3k_B$. The typical Debye screening length for the electrons, $\lambda_D = \sqrt{\varepsilon_0 k_B T_e/n_e e^2}$, is from $1-10$ μm.

Ions, on the other hand, inherit the low temperature of neutral atoms trapped in the MOT. However, the ions rapidly heat after plasma formation due to the excess potential energy in their random spatial distribution [13, 21]. This phenomena has been called "disorder induced heating" or "correlation-heating", and it was first predicted theoretically in [22]. Due to this phenomenon, after 1 μs the ions will have a temperature on the order of 1 K. A more detailed study of this effect follows in the last section.

After this initial ion heating, the electron thermal pressure causes the entire plasma to expand into the surrounding vacuum. This was studied theoretically in [23, 24, 22, 25, 15] and was seen experimentally in [26, 27, 21, 13, 28, 29]. The expansion is self-similar and the initial Gaussian profile is preserved, as long as the initial electron Coulomb coupling parameter is smaller than 1. This condition is violated at very high initial plasma density and very low initial electron temperature [18].

For a self-similar plasma expansion, the velocity profile is given by [28]

$$\mathbf{u}(\mathbf{r}) = \gamma(t)\mathbf{r},$$
$$\gamma(t) = \frac{t/\tau_{exp}^2}{1+t^2/\tau_{exp}^2}, \quad (1)$$

where $\tau_{exp} = \sqrt{m_i \sigma(0)^2 / k_B [T_e(0) + T_i(0)]}$ is the characteristic expansion time set by initial plasma parameters, and $\mathbf{r}$ is the position vector relative to plasma center. $\sigma(t)$ is the time-varying $1/e^2$ radius of plasma and $T_\alpha(t)$ is the temperature for each plasma component. The inset of Fig. 5 shows a measurement of the expansion energy of the whole plasma cloud as it increases with respect to time. A more detailed discussion on the dynamics of ultracold neutral plasmas can be found in a recent review article [2].

FLUORESCENCE SPECTROSCOPY AND ANNULAR ANALYSIS

To characterize and measure the dynamics of UNPs, we use fluorescence spectroscopy and annular analysis. Figure 3 shows a schematic of the fluorescence imaging probe. The plasma is illuminated along $\hat{y}$ with a laser beam that is nearly resonant with the $^2S_{1/2} - ^2P_{1/2}$ transition in Sr$^+$ at $\lambda = 422$ nm. The fluorescence excitation beam is spatially narrowed into a sheet, allowing for localized analysis of ion temperatures within a volume of the plasma with small density variation. Fluorescence is recorded by the camera in a direction perpendicular to the probe beam ($-\hat{z}$). The intensity and duration of the fluorescence beam is low enough so the ions typically scatter fewer than one photon during the time the beam is on, preventing any optical pumping to the metastable $^2D_{3/2}$ state.

The spatially-resolved fluorescence signal at laser frequency, v, can be mathematically described by [28]

$$F(v,x,y) = C\frac{\gamma_0}{2}\int dz \rho_i(\mathbf{r})\frac{I(\mathbf{r})}{I_{sat}}\int ds \frac{\gamma_0/\gamma_{eff}}{1+\left[\frac{2(v-s)}{\gamma_{eff}/2\pi}\right]^2} \times \quad (2)$$

$$\frac{1}{\sqrt{2\pi}\sigma_D[T_i(\mathbf{r})]}\exp\left\{-\frac{[s-(v_0+v_{exp}^y(\mathbf{r}))]^2}{2\sigma_D^2[T_i(\mathbf{r})]}\right\}.$$

The multiplicative factor, C, accounts for the collection solid angle, dipole radiation pattern orientation, and detector efficiency. $\rho_i(\mathbf{r})$ is the ion density and $\gamma_{eff} = \gamma_0 + \gamma_{ins}$

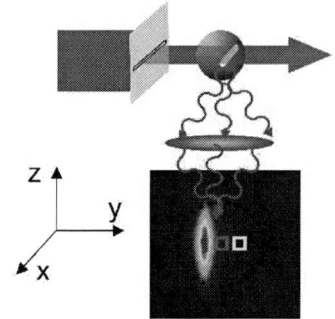

FIGURE 3. Fluorescence probe for UNPs using light resonant with the main transition of Sr^+. The laser beam propagates perpendicularly to a CCD intensified camera and the imaging axis. The fluorescence excitation beam is spatially narrowed into a sheet. As such, the beam interacts with a small volume of the plasma where density variation is small. A spectrum can be constructed from a series of images, each taken at different laser beam frequencies. Furthermore, small regions of the images can be analyzed (highlighted square regions shown inside the plasma outline) with this imaging probe. The fluorescence geometry has the advantage of decoupling the effects of expansion energy and thermal energy within this regional analysis. From spectroscopy, plasma parameters such as plasma size, ion velocity and temperature are extracted.

is the effective Lorentzian linewidth due to the natural linewidth of the transition, $\gamma_0 = 2\pi \times 20\,\text{MHz}$, and any instrumental linewidth, γ_{ins}. The intensity profile of the fluorescence excitation beam is $I(\mathbf{r})$ and we assume $I(\mathbf{r}) \ll I_{sat}$. The saturation intensity with linearly-polarized light is $I_{sat} = 114\,\text{mW/cm}^2$, where the Clebsch-Gordon coefficients for the transition have been taken into account . The conventional Doppler width is $\sigma_D(T) = \sqrt{k_B T_i(\mathbf{r})/m_i}/\lambda$.

The average resonance frequency of the transition for ions is Doppler-shifted from the resonance frequency, v_0, due to plasma expansion. The shift is given by $v_{exp}^y(\mathbf{r}) = \mathbf{u}(\mathbf{r}) \cdot \hat{y}/\lambda$. For a self-similar expansion, (1) yields

$$v_{exp}^y(\mathbf{r}) = \gamma(t)y/\lambda, \qquad (3)$$

which produces a correlation between position in the camera and the Doppler shift due to expansion. Figure 4 demonstrates how the correlation of position and Doppler shift make the expansion appear in the spectra as a shift of the central frequency. On the other hand, the Doppler width is only representative of thermal motion.

Figure 5 shows fluorescence measurements of two regions of the plasma as well as measurements using the whole cloud image for analysis. Notice that in the whole cloud measurements the expansion energy dominates over the thermal energy with values an order of magnitude higher than the thermal energy of the individual regions.

Regional analysis of fluorescence imaging has the advantage of decoupling the effect of thermal energy and thermal energy in the spectra. On the other hand, the regions are usually small, $\sim 0.1\sigma(t)$, which results in a small signal-to-noise ratio. One could increase the size of the regions to increase the signal; however, as the regions become bigger so does the contribution of the expansion energy to the broadening of the spectra.

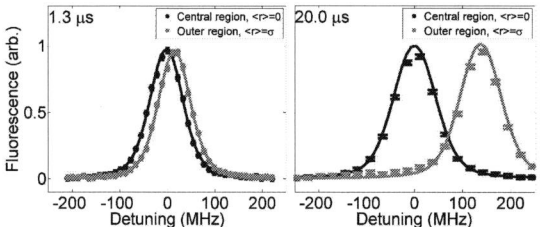

FIGURE 4. Fluorescence spectra of two restricted regions of the plasma cloud showing the separation of expansion and random thermal motion effects. The plots show both regional spectra at different times after plasma formation (time shown on the upper left corner). From the width of the spectra, one can extract a measurement of the ion thermal energy in that region of the plasma. The outer region with $\langle r \rangle \neq 0$ is Doppler shifted due to the plasma expansion.

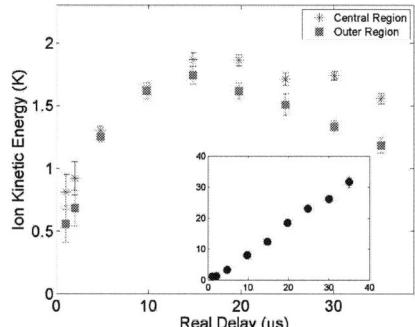

FIGURE 5. Ion kinetic energy measurements done with regional analysis of fluorescence imaging, which permits the decoupling of the expansion and thermal energies. Both central and outer region measurements show the thermal energy of that particular region. Due to disorder induced heating, the initial kinetic energy of the central region is higher than that of the outer region. The inset shows fluorescence imaging measurements done with whole cloud analysis where the expansion effect dominates because it cannot be decoupled. The initial conditions for these measurements are $T_e(0) = 48$ K, the peak ion density is 9×10^{14} m^{-3} and $\sigma(0) = 1.9$ mm.

To increase the signal-to-noise ratio without including expansion energy, one needs to take advantage of the symmetry of the system. Ultracold Neutral Plasmas have spherically-symmetric density distributions. Regions of the plasma with the same radial distance from the center have similar dynamics. As a consequence, one could, in principle, combine all regions at constant radius to an annulus. We call this procedure annular analysis of the fluorescence images.

One complication to this analysis is that, as can be seen from (3), different sections of an annulus have different Doppler shifts due to the expansion. Adding the signal from sections with different $\langle y \rangle$ positions and resulting different frequency shifts would then broaden the total spectrum of the annulus. In order to counter this effect and be able to sum the contribution from distinct sections of an annulus, one has to remove the

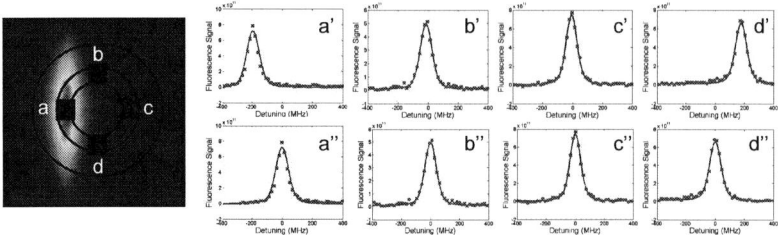

FIGURE 6. Removal of the Doppler shift from distinct sections of an annulus. The advantage of fluorescence regional analysis is that it decouples the effect of expansion and thermal energies. Shown above are four regions of an annulus (a,b,c,d) with the respective spectrum (a',b',c',d'). Notice that b' and c' have no frequency shift since $\langle y \rangle = 0$ for these sections. In the final set of spectra (a'',b'',c'',d'') the Doppler shift from each region has been subtracted. Upon removal of the Doppler shift, the signal from each of these regions can be added together.

frequency shift from each individual position. This is only possible because the expansion velocity profile is well known (1). The process of removing the Doppler shift from expansion is demonstrated in Fig. 6. As a result of adding the signal from symmetrically-similar regions, signal-to-noise ratio is higher than with simple regional analysis. This allows regions with negligible density variation to be analyzed. An example of such analysis is shown in the next section.

DISORDER INDUCED HEATING AND KINETIC ENERGY OSCILLATIONS

As mentioned in the previous section, annular analysis permits the study of regions with insignificant density variation. This is beneficial for studying density-dependent phenomena such as disorder induced heating and kinetic energy oscillations.

Figure 7 shows a kinetic energy measurement performed with annular analysis of sheet fluorescence images. The ions inherit the spatially-uncorrelated state of the neutral atoms, resulting in a large amount of potential energy. Ions equilibrate and develop correlations; their kinetic energy increases. This process is called disorder induced heating, or correlation heating. The typical ion energy after equilibration is about $e^2/4\pi\varepsilon_0 a$, which is on the order of 1 K of energy. A more accurate expression was derived in [22] taking into account the screening effects of the electrons. This agrees well with the measurement shown and with previous experiments [30].

The time-scale for disorder induced heating corresponds to the inverse of the ion plasma oscillation frequency, $1/\omega_{pi} = \sqrt{m_i\varepsilon_0/n_i e^2}$, which is on the order of a few hundred nanoseconds. As can be seen from the formula, this oscillation frequency is density-dependent, making annular analysis of sheet fluorescence an ideal tool to study this phenomenon.

Figure 7 shows oscillations of the kinetic energy as the ions undergo harmonic motion in the potential cage created by their neighbors. The kinetic energy later settles down to its equilibrium value. This oscillatory picture is only appropriate in a system where the

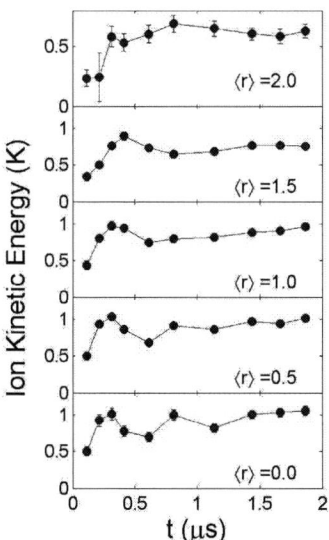

FIGURE 7. Annular analysis of kinetic energy oscillations in a UNP. As spatial correlations develop, ion potential energy is converted to kinetic energy. The kinetic energy oscillates about its equilibrium value with the plasma frequency, $\omega_{pi} \propto n^{1/2}$. A UNP has a Gaussian density distribution, therefore, the central annulus $\langle r \rangle = 0$ has higher density than any other annulii and oscillates faster. Notice how the oscillation frequency and the equilibrium energy value diminishes as $\langle r \rangle$ increases.

correlation energy between particles is greater than the thermal energy, namely strongly-coupled plasmas such as UNPs. Previous experiments used absorption imaging to perform these measurements [30, 31]. As can be seen from Fig. 1, absorption imaging beam probes the whole plasma cloud along $-\hat{z}$, averaging over the density distribution. This averaging leads to a reduced contrast in the oscillation, making it difficult to study the damping rate of such oscillations. The damping has been studied for a one-component plasma [32] and was predicted to depend sensitively on the Coulomb coupling parameter. Contrary to absorption imaging, annular analysis of sheet fluorescence does not have density averaging and leads to much higher contrast oscillations as shown in Fig. 7. This type of measurement should allow further study of the underlying damping rate of the oscillations.

A non-invasive probe that permits measurements with small density variation is crucial in investigating any phenomenon. Strong-coupling effects in ultracold neutral plasmas, such as disorder induced heating and kinetic energy oscillations [22, 32], have a strong dependence on density and benefit from these types of measurements. Furthermore, the development of annular analysis of fluorescence images will allow us to probe the temperature of UNPs and to study ion adiabatic cooling and electron-ion thermalization.

ACKNOWLEDGMENTS

This research was supported by the Department of Energy Office of Fusion Energy Sciences, the National Science Foundation, and the David and Lucille Packard Foundation.

REFERENCES

1. T. C. Killian, *Science* **316**, 705 (2007).
2. T. C. Killian, T. Pattard, T. Pohl, and J. M. Rost, *Phys. Rep.* **449**, 77 (2007).
3. S. Ichimuru, *Rev. Mod. Phys.* **54**, 1017 (1982).
4. H. M. V. Horn, *Science* **252**, 384 (1991).
5. M. Nantel, G. Ma, S. Gu, C. Y. Cote, J. Itatani, and D. Umstadter, *Phys. Rev. Lett.* **80**, 4442 (1998).
6. E. Springate, N. Hay, J. W. G. Tisch, M. B. Mason, T. Ditmire, M. H. R. Hutchinson, and J. P. Marangos, *Phys. Rev. A* **61**, 063201 (2000).
7. G. E. Morfill, H. M. Thomas, U. Konopka, and M. Zuzic, *Phys. Plasmas* **6**, 1769 (1999).
8. T. B. Mitchell, J. J. Bollinger, X. P. Huang, W. M. Itano, and D. H. E. Dubin, *Phys. Plasmas* **6**, 1751 (1999).
9. M. D. Perry, and G. Mourou, *Science* **264**, 917 (1994).
10. J. Lindl, *Phys. Plasmas* **2**, 3933 (1995).
11. E. L. Clark, K. Krushelnick, M. Zepf, F. N. Beg, M. Tatarakis, A. Machacek, M. I. K. Santala, I. Watts, P. A. Norreys, and A. E. Dangor, *Phys. Rev. Lett.* **85**, 1654 (2000).
12. R. A. Snavely, M. H. Key, S. P. Hatchett, T. E. Cowan, M. Roth, T. W. Phillips, M. A. Stoyer, E. A. Henry, T. C. Sangster, M. S. Singh, S. C. Wilks, A. MacKinnon, A. Offenberger, D. M. Pennington, K. Yasuike, A. B. Langdon, B. F. Lasinski, J. Johnson, M. D. Perry, and E. M. Campbell, *Phys. Rev. Lett.* **85**, 2945 (2000).
13. C. E. Simien, Y. C. Chen, P. Gupta, S. Laha, Y. N. Martinez, P. G. Mickelson, S. B. Nagel, and T. C. Killian, *Phys. Rev. Lett.* **92**, 143001 (2004).
14. D. O. Gericke, and M. S. Murillo, *Contrib. Plasma Phys.* **43**, 298 (2003).
15. F. Robicheaux, and J. D. Hanson, *Phys. Plasmas* **10**, 2217 (2003).
16. S. G. Kuzmin, and T. M. O'Neil, *Phys. Plasmas* **9**, 3743 (2002).
17. S. Mazevet, L. A. Collins, and J. D. Kress, *Phys. Rev. Lett.* **88**, 55001 (2002).
18. P. Gupta, S. Laha, C. E. Simien, H. Gao, J. Castro, T. C. Killian, and T. Pohl, *Phys. Rev. Lett.* **99**, 75005 (2007).
19. H. J. Metcalf, and P. van der Straten, *Laser Cooling and Trapping*, Springer-Verlag, New York, New York, 1999.
20. S. B. Nagel, C. E. Simien, S. Laha, P. Gupta, V. S. Ashoka, and T. C. Killian, *Phys. Rev. A* **67**, 011401 (2003).
21. E. A. Cummings, J. E. Daily, D. S. Durfee, and S. D. Bergeson, *Phys. Rev. Lett.* **95**, 235001 (2005).
22. M. S. Murillo, *Phys. Rev. Lett.* **87**, 115003 (2001).
23. T. Pohl, T. Pattard, and J. M. Rost, *Phys. Rev. A* **70**, 033416 (2004).
24. T. Pohl, T. Pattard, and J. M. Rost, *Phys. Rev. Lett.* **92**, 155003 (2004).
25. F. Robicheaux, and J. D. Hanson, *Phys. Rev. Lett.* **88**, 55002 (2002).
26. S. Kulin, T. C. Killian, S. D. Bergeson, and S. L. Rolston, *Phys. Rev. Lett.* **85**, 318 (2000).
27. E. A. Cummings, J. E. Daily, D. S. Durfee, and S. D. Bergeson, *Phys. Plasmas* **12**, 123501 (2005).
28. S. Laha, P. Gupta, C. E. Simien, H. Gao, J. Castro, and T. C. Killian, *Phys. Rev. Lett.* **99**, 155001 (2007).
29. T. C. Killian, Y. C. Chen, P. Gupta, S. Laha, Y. N. Martinez, P. G. Mickelson, S. B. Nagel, A. D. Saenz, and C. E. Simien, *J. Phys. B: At. Mol. Opt. Phys.* **38**, 351 (2005), (Equations 7, 10, 11, and 17 should be multiplied by γ_0/γ_{eff}.).
30. Y. C. Chen, C. E. Simien, S. Laha, P. Gupta, Y. N. Martinez, P. G. Mickelson, S. B. Nagel, and T. C. Killian, *Phys. Rev. Lett.* **93**, 265003 (2004).
31. S. Laha, Y. C. Chen, P. Gupta, C. E. Simien, Y. N. Martinez, P. G. Mickelson, S. B. Nagel, and T. C. Killian, *Euro. Phys. J. D* **40**, 51 (2006).
32. G. Zwicknagel, *Contrib. Plasma Phys.* **39**, 155 (1999).

Using Charged Particle Imaging to Study Ultracold Plasma Expansion

X. L. Zhang, R. S. Fletcher, and S. L. Rolston

Joint Quantum Institute, Department of Physics, University of Maryland, College Park, MD 20742

Abstract.
We develop a projection imaging technique to study ultracold plasma dynamics. We image the charged particle spatial distributions by extraction with a high-voltage pulse onto a position-sensitive detector. Measuring the 2D width of the ion image at later times (the ion image size in the first 20 μs is dominated by the Coulomb explosion of the dense ion cloud), we extract the plasma expansion velocity. These velocities at different initial electron temperatures match earlier results obtained by measuring the plasma oscillation frequency. The electron image size slowly decreases during the plasma lifetime because of the strong Coulomb force of the ion cloud on the electrons, electron loss and Coulomb explosion effects.

Keywords: ultracold plasma, plasma expansion, charged particle imaging
PACS: 52.27.Aj,52.27.Gr,52.70.-m

Ultracold plasmas (UCPs), formed by photoionizing laser-cooled atoms near the ionization limit, have well controlled initial conditions and slow dynamics compared to other laser-produced plasma systems, and thus provide a clean and simple source with excellent spatial and temporal resolution available to study basic plasma phenomena. In the majority of experiments to date, UCPs have been unconfined and freely expanded into vacuum, a fundamentally important dynamic in laser-produced plasmas as well as UCPs. The first experimental study of the expansion of UCPs was performed using the plasma frequency as a probe of the plasma density as a function of time [1]. By applying a small RF electric field to excite plasma oscillations, the plasma density time dependence was mapped from the oscillations, and a ballistic expansion of the plasma was found, i.e. $\sigma^2(t) = \sigma_0^2 + v_0^2 t^2$. For initial electron temperatures $T_e(0) \geq 70$ K, the expansion velocities follow $v_0^2 = k_B T_e/m_i$, the ion acoustic velocity due to the electron pressure on the ions, in agreement with a simple hydrodynamics model. At low initial temperatures, the UCPs expand faster than expected, which indicates plasma heating. The expansion dynamics of UCPs have subsequently been studied experimentally by various methods, such as plasma collective modes [2], absorption imaging [3], fluorescence imaging [4], and theoretically [5, 6, 7, 8].

In this work we use a projection imaging technique to study the UCP dynamics during the full lifetime of the plasma. We image the charged particle (ions or electrons) spatial distribution of an expanding UCP by extracting them with a high-voltage pulse and accelerating them onto a position-sensitive detector. The expansion is self-similar, as the ion (or electron) cloud maintains a Gaussian density profile throughout the lifetime of the plasma. Early in the lifetime of the plasma, the ion image size is dominated by the Coulomb explosion of the dense ion cloud. The image size is at a minimum at about 20 μs and then afterwards increases (the Coulomb explosion of the ion cloud becomes negligible), reflecting the true size of the plasma. We obtain the ion image width by 2D Gaussian fitting, and extract the final asymptotic expansion velocity by fitting the linear region of the ion images as a function of time at later plasma times (≥ 20 μs). Assuming that the ion cloud maintains the Gaussian density distribution during the Coulomb explosion phase, we can extract the actual ion cloud size from the ion projection image by accounting for the Coulumb explosion effect. The plasma size indeed follows the ballistic expansion as expected from a simple hydrodynamics model throughout the whole lifetime of the plasma. Including the corrected plasma sizes in the first 20 μs compared to that obtained by only fitting the linear region at later times only amounts to a few percent change in asymptotic velocity. The velocities at different initial electron temperatures match earlier results obtained by measuring the plasma oscillation frequency [1], which provides strong support for this method to study the UCP expansion and the previous technique. We can also image the electrons during the lifetime of an UCP by switching the polarity of the high-voltage pulse and the accelerating voltages on the grids. Unlike the ion images, the size of the electron image slowly decreases during the lifetime of UCP because of the strong Coulomb effect of the dense ion cloud on the electrons, electron loss and the Coulomb explosion effect. This technique provides a good tool to study UCP dynamics in a magnetic field, such as expansion[9] and plasma instabilities[10].

Details of the creation of ultracold neutral plasmas are described in [11]. We accumulate about 10^6 metastable Xenon atoms, trapped and cooled in a magneto-optical trap to a temperature of ~ 20 μK. The atomic cloud has a Gaussian spatial distribution with a peak density of about 2 x 10^{10} cm^{-3}, and an rms-radius of ~ 0.3 mm. We produce the plasma with a two-photon excitation process (882-nm photon from the cooling laser and 514-nm photon from a 10-ns pulsed dye laser), ionizing up to 30% of the atoms. We control the ionization fraction with the intensity of the photoionization laser, while the initial electron energy is controlled by tuning the 514-nm photon energy with respect to the ionization limit, usually in the 1-1000 K range. The ionized cloud rapidly loses a few percent of the electrons, resulting in a slightly attractive potential for the remaining electrons, and quickly reaches a quasineutral plasma state. It then freely expands with an asymptotic velocity v_0 typically in the 50-100 m/s range caused by the outward electron pressure[1].

For projection imaging of charged particles, external electric fields are applied via four grids to direct and accelerate them towards a position-sensitive detector (a microchannel plate detector with phosphor screen) (figure 1b). Two grids ("top" and "bottom" grids) are 1.5 cm above and below the plasma, and the other two ("middle" and "front" grids) are located between the bottom grid and the detector. By applying a high-voltage pulse to the top grid at a specific delay time after the formation of the UCP and with accelerating voltages on the middle and front grids (-300 V and -700 V for ions, 130 V and 300 V for electrons), we image the charged particle distribution of expanding UCPs onto the phosphor screen. The phosphor image, recorded by a CCD camera, is proportional to the charged particle density, and weakly sensitive to their energy. The high-voltage pulse has an amplitude of 340 V for ions (-200 V for electrons), a width of 4 μs, and a rise time of 60 ns. It is generated by modifying the square pulse generator used in ion beam deflection in a neutron generator [12], which uses power FETs to fast-switch a HV source. Figure 2 shows typical ion projection images (2D ion spatial

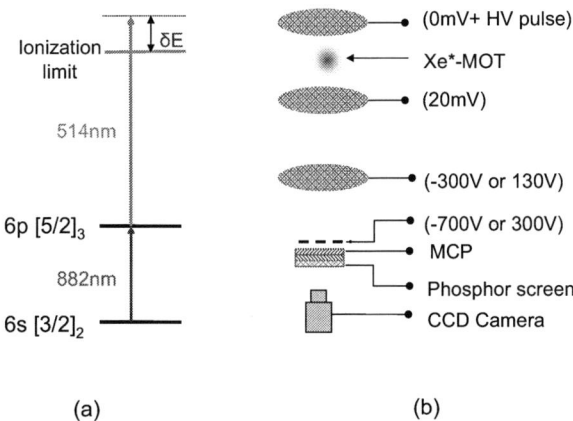

FIGURE 1. Two-photon excitation process and experimental setup. (a) two-photon excitation process: one photon (red solid line) at 882 nm drives the 6s[3/2]$_2$ → 6p[5/2]$_3$ transition, and the other (10-ns pulse) at 514 nm ionizes the atoms in the 6p[5/2]$_3$ state. (b) experimental setup for imaging the charged particles onto the MCP/phosphor screen.

distributions integrated over the third dimension) with averages of 8 images to increase the signal-to-noise ratio. The units are in pixels, which correspond to approximately 150 μm. Figure 2a is a false color plot of an ion image of an UCP at a 20 μs delay time and initial electron temperature $T_e(0)$ of 100 K, which fits well to a 2D Gaussian profile (figure 2b). The ion images maintain a Gaussian profile during most of the lifetime of the UCP as shown in figure 2c-2f. We note that the ion profiles have a flat top and even a dip at very late times of about 150-200 μs (figs. 2d and 2f, especially for high $T_e(0)$), and this appears earlier for higher $T_e(0)$. It is currently unknown what causes the flat top and dip in the center of the ion images.

We extract the plasma size by 2D Gaussian fitting of the ion images (figs. 2c and 2e) at specific delay times after the formation of the UCP. Figure 3 shows the measured plasma sizes as a function of delay time for different $T_e(0)$. The curves are for $T_e(0)$ of 400 K, 200 K, 60 K and 10 K from top to bottom respectively. Early in the lifetime of the plasma, the measured image size is dominated by the strong Coulomb explosion of the dense ion cloud during transit to the detector, not the true size of the plasma. This is because the electrons are extracted from the UCPs very quickly (a few ns) by the HV pulse, leaving a Gaussian distribution of charged ions. The ions then fly to the detector in about 9 μs, set by the HV pulse and the accelerating voltages of the other grids (the time-of-flight time of the ions to the detector can be determined from the delay time of the ion current after the formation of the plasma relative to the HV pulse). At early times ($\leq$ 20 μs), the plasma size is still small (on the order of the intial size, several hundred microns) and the strong Coulomb repulsion between the ions produces a large ion image. As the plasma size increases, the Coulomb explosion effect diminishes and

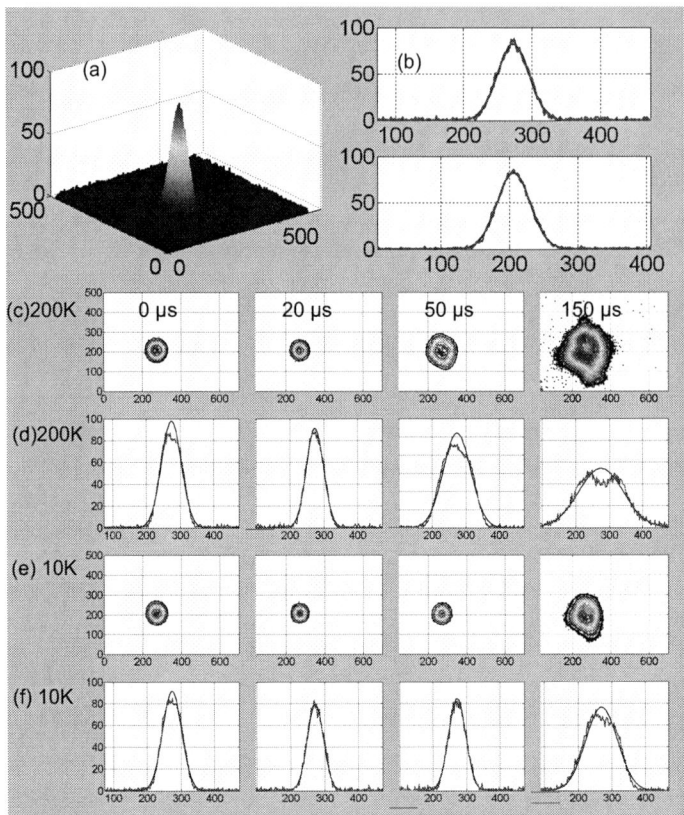

FIGURE 2. (a) a false color ion image (2D ion spatial distribution integrated over the third dimension) of an expanding UCP at t = 20 μs, $T_e(0)$ = 100 K; (b) the 2-D Gaussian fits (blue curves) of the ion image (a) along the x and y axis in the horizontal plane (red curves); (c) and (e) are the contour plots of the ion images at different delay times for $T_e(0)$ = 200 K and 10 K, respectively; (d) and (f) are the corresponding 2-D Gaussian fits of (c) and (e). All the size related units in (a)-(f) are in pixels, and one pixel corresponds to 150 μm. The y axis of (d) and (f) is in arbitrary units.

no longer affects the measured size. The image size is at a minimum at about 20 μs and afterwards increases, reflecting the true size of the plasma with a constant magnification factor of 1.3 (discussed below), as expected from the ballistic expansion model. The size increases slowly and the minimum point of the measured plasma size moves to a later time as we decrease $T_e(0)$, because the expansion velocity which depends on $T_e(0)$ gets smaller, and also the Coulomb explosion effect diminishes more slowly because of the slower expansion.

Assuming that the ion distribution is not affected by the fast HV pulse and maintains a Gaussian distribution during the ion transport to the detector, we can extract the initial ion cloud size from the ion projection image by correcting for the Coulomb explosion

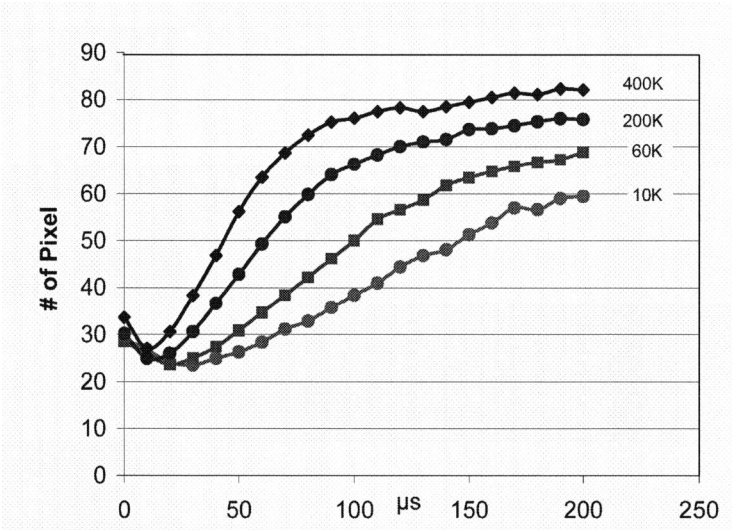

FIGURE 3. Measured plasma size as a function of elapsed time after the formation of the UCP for different $T_e(0)$. The curves are for $T_e(0)$ of 400 K, 200 K, 60 K and 10 K from top to bottom, respectively. Early in the lifetime of the plasma, the size of the image is dominated by the Coulomb explosion of the dense ion cloud.

effect. This is done as follows: First, we start with the time dependent plasma density distribution $n(r,t) = n_0 (\sigma_0/\sigma_t)^3 e^{-r^2/2\sigma_t^2}$; then, we calculate the Coulomb potential of the ion cloud at the specific delay time and extract the average acceleration of the ion cloud; next, we obtain the ion cloud size and expansion velocity after the Coulomb explosion with a small time-of-flight step (small enough for constant acceleration for each iteration); finally, we iterate this procedure to get the final ion cloud size after the total time-of-flight, which agrees with the measured ion size. That is, the plasma size indeed follows the ballistic expansion as expected from a simple hydrodynamics model throughout the whole lifetime of the UCP. This only results in a few percent change in the plasma expansion velocity by including the corrected plasma sizes of the early times compared to that found by only fitting the linear region of the ion image sizes at later times. If we also consider that the ion cloud will freely expand with the ion acoustic velocity in addition to the Coulomb explosion during the time-of-flight, we need to shift the plasma sizes up by several hundred micrometers, which is equivalent to shifting the x-axis (time) in figure 3, but this does not affect the extracted plasma expansion velocity. At later times, especially for high $T_e(0)$, the measured size does not linearly increase. This is partly because the size of the UCP is large enough to be affected by the 4 posts that secure the grids above and below the plasma (the top and bottom grids), and it approaches the 3-cm size of the detector.

By fitting to the sizes after about 20 μs (for high $T_e(0)$, (only fitting the restricted linear region), we can get the asymptotic expansion velocities of UCPs at different $T_e(0)$ with a magnification factor of 1.3 due to an ion lensing effect (the red solid curve with square points in figure 4). The magnification arises from the electric fields which tend to focus or expand the ions (depending on the voltage settings of the grids) as they transport to the detector. It is confirmed by adjusting the voltages on the grids (especially the middle and front grids), which strongly affects the ion image size as well as the scaling factor. By using an ion optics simulation program, we simulate our ion projection imaging setup with the actual spacings and voltage settings of the grids, and find the ion magnification factor from the trajectories (which is 1.3 for the images at figure 2). At high $T_e(0)$ ($\geq 60K$), the expansion velocities $v_0 \propto \sqrt{T_e}$ as expected from a simple hydrodynamics model, that is, the slope of the red dashed line in figure 4 is about 1/2; at lower $T_e(0)$ ($\leq 60K$), the expansion velocities are higher than expected from the self-similar expansion, which indicates heating. The black solid line with open circles is the asymptotic expansion velocity obtained by measuring the plasma oscillation frequency [1]. The good agreement between our experimental results and earlier results obtained by measuring the plasma oscillation frequency strongly supports the measurement of UCP expansion velocities with both the projection imaging method and the previous technique.

The excess expansion velocity at low T_e as seen in [1] and in fig. 4 is attributable to heating due to three-body recombination (TBR) collisions, and was verified in [13], which directly observed the Rydberg atoms formed in the collisions. The $T_e^{-9/2}$ dependence of TBR makes it important in UCPs, and especially at low T_e. Using the same imaging technique as for ions, we can also image the electrons by reversing the polarity of the HV pulse and the voltages on the grids between the plasma and detector. Figure 5 is the measured electron size as a function of elapsed time after the formation of the UCP. The black curve with circles is the measured electron size extracted from the 2D Gaussian fitting of the images. The brown curve with triangles is the electron size with the actual scaling factor due to an electron lensing effect, which is consistent with a theoretical calculation of electron size with a 100-ns Coulomb explosion time (the red curve with open squares), which also took into account electron loss due to the evaporation of electrons out of the system. We assume that the ion cloud follows a ballistic expansion model (the blue curve with diamonds), while the electron distribution is initially identical to the ion distribution, but with a truncation at the appropriate radius such that the total electron number agrees with the measured charge imbalance at a specific delay time. We then perform self-consistent calculations for the plasma potential to extract the final electron size (the magenta curve with square points). The electron magnification is obtained from trajectory simulations with the voltage settings of the grids and the ion spatial distribution. We note that the electron lensing factor, unlike that of the ions, is not constant during the whole UCP lifetime due to the strong Coulomb force of the ion cloud on the much lighter electrons, especially for the first 30 μs of the plasma lifetime. The electrons are removed from the plasma in a few ns after applying the HV pulse, but the ions maintain their Gaussian spatial distribution during that short period of time, which

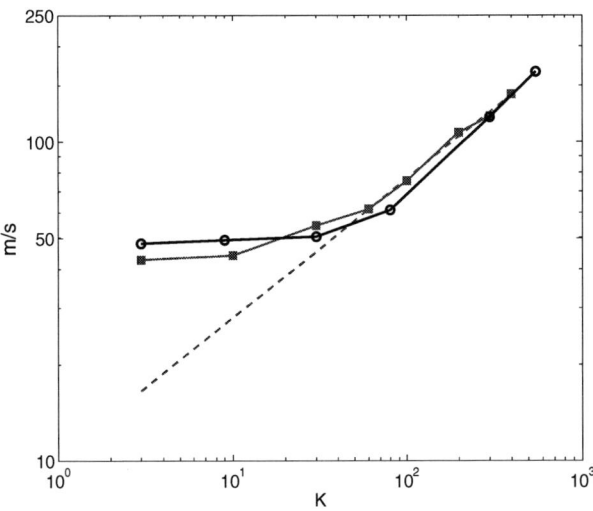

FIGURE 4. The asymptotic expansion velocity as a function of initial electron temperature T_e. The red solid line with square points is the experimental result which matches the results obtained by measuring the plasma oscillation frequency (the black solid curve with circle points)[1]. The red dashed line is the linear fitting of the data above 60 K with a slope of 1/2.

exerts a strong Coulomb force on the electrons and partially cancels the applied electric field. This increases the electron lensing, confirmed by the trajectory simulation. As the plasma expands, the plasma size gets larger, and the Coulomb force on the electrons due to the ion cloud gets smaller, so the electron lensing tends towards a constant at later plasma times (≥ 30 μs). The ion lensing is constant because there are no electrons left during the ion Coulumb explosion phase.

In conclusion, we have developed a projection imaging technique to study the dynamics of an expanding ultracold plasma. Unlike the previous experimental technique that used the plasma oscillation frequency, which only worked at early times, this method can study expansion dynamics for the entire plasma lifetime. In addition, we can measure the both the evolving electron and ion spatial distributions. This method will be useful for further study of ultracold plasmas, including plasma instabilities, and plasma expansion under different conditions of magnetic confinement.

ACKNOWLEDGMENTS

This work was partially supported by the National Science Foundation PHY-0714381.

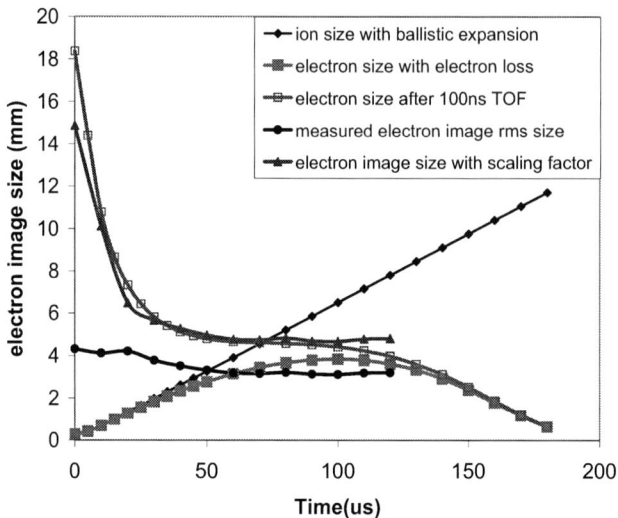

FIGURE 5. The electron sizes of UCPs as a function of time for $T_e(0) = 100$ K. The black curve with dots is the electron size extracted from the 2D Gaussian fitting of the electron images. The brown curve with triangles is the electron size with the a scaling factor due to the charged particle lensing effect, which is consistent with the theoretical calculated electron size with a 100 ns Coulomb explosion time (the red curve with unfilled squares).

REFERENCES

1. S. Kulin, T. C. Killian, S. D. Bergeson, and S. L. Rolston, *Phys. Rev. Lett.* **85**, 318 (2000).
2. R. S. Fletcher, X. L. Zhang, and S. L. Rolston, *Phys. Rev. Lett.* **96**, 105003 (2006).
3. C. E. Simien, Y. C. Chen, P. Gupta, Y. N. Martinez, P. G. Mickelson, S. B. Nagel, and T. C. Killian, *Phys. Rev. Lett.* **92**, 14 (2004).
4. E. A. Cummings, J. E. Daily, D. S. Durfee, and S. D. Bergeson, *Phys. Rev. Lett.* **83**, 235001 (2005).
5. S. D. Bergeson, and R. L. Spencer, *Phys. Rev. E* **67**, 026414 (2003).
6. F. Robicheaux, and J. D. Hanson, *Phys. Plasmas* **10**, 2217 (2003).
7. T. Pohl, T. Pattard, and J. Rost, *Phys. Rev. Lett.* **92**, 15 (2004).
8. S. Mazevet, L. Collins, and J. D. Kress, *Phys. Rev. Lett.* **88**, 5 (2002).
9. X. L. Zhang, R. S. Fletcher, and S. L. Rolston, *Phys. Rev. Lett.* **100**, 253002 (2008).
10. X. L. Zhang, R. S. Fletcher, and S. L. Rolston, Observation of an ultracold plasma instability, arXiv-0806.4691.
11. T. C. Killian, S. Kulin, S. D. Bergeson, L. A. Orozco, C. Orzel, and S. L. Rolston, *Phys. Rev. Lett.* **83**, 4776 (1999).
12. D. Tomic, *Rev. Sci. Instrum.* **61**, 1729 (1990).
13. T. C. Killian, M. L. Lim, S. Kulin, R. Dumke, S. D. Bergeson, and S. L. Rolston, *Phys. Rev. Lett.* **86**, 3759 (2001).

Observation of String Ion Cloud in a Linear RF Trap

M. Aramaki[a], S. Kameyama[a], Y. Sakawa[b], T. Shoji[c], and A. Kono[a]

[a]Department of Electrical Engineering and Computer Science, Nagoya University,
Furo-cho, Chikusa-ku, Nagoya, 464-8603, Japan
[b]Institute of Laser Engineering, Osaka University,
Yamadaoka, Suita, Osaka, 565-0871, Japan
[c]Department of Energy Engineering and Science, Nagoya University,
Furo-cho, Chikusa-ku, Nagoya, 464-8603, Japan

Abstract. We aim to study the effect of the long-range correlation among ions on their statistical characteristics using ion clouds confined in a linear rf ion trap. It is important to keep the ion cloud in one dimension, where the influence of the rf heating is negligible, for the detailed research on the effect of the Coulomb interaction on the statistical characteristics of the ion cloud. In this paper, the method of the generation of an ideal ion string is proposed. We also briefly report the performances of our experimental equipment and the preliminary results of generation of ideal 1D ion cloud.

Keywords: rf trap, laser cooling, non-Boltzmann-Gibbs statistics
PACS: 50

INTRODUCTION

Boltzmann-Gibbs statistics have achieved a great success in describing the thermal equilibrium state of short-range interacting systems. Gaussian distribution that is derived as the maximum entropy state of the Boltzmann-Shannon entropy is essentially common in nature. However, the validity of Boltzmann-Gibbs statistics for systems with long-range interaction or long-time memory is not clear. Recently, the several types of non- Boltzmann-Gibbs statistics have been proposed. The Tsallis entropy is one of the possible generalizations of the Boltzmann-Shannon entropy, which is defined as

$$S_q[p] = \frac{k}{1-q}\left[\sum_{i=1}^{W}(p_i)^q - 1\right],$$

where k is the Boltzmann constant, p_i is the probability associated with a event, and W is the total number of possible configurations [1]. The nonextensibility of the Tsallis entropy implies the suitability for the system governed by long-range interaction or long-time memory. Many theoretical and experimental researches on the Tsallis statistics have been reported [2], however there is no clear explanation how q can be determined in individual experimental systems. We aim to study the effect of the long-range correlation among ions on their statistical characteristics using ion clouds confined in a linear rf ion trap.

Coulomb systems composed of ions trapped in an rf trap is suitable for the detailed research on long-range correlating systems, since the interaction between the ions and surrounding environment is weak, and the scale of the system can be easily controlled. The correlation among the trapped ions is indicated by the Coulomb coupling parameter Γ defined as the ratio of the Coulomb interaction between ions to the thermal energy of ions:

$$\Gamma = \frac{1}{4\pi\varepsilon_0}\frac{Q^2}{a_{ws}k_B T}.$$

Here, ε_0 is the permittivity of the vacuum, Q is the charge of an ion, k_B is the Boltzmann's constant, T is the temperature, a_{ws} is the Wigner-Seitz radius:

$$a_{ws} = \left(\frac{3}{4\pi n}\right)^{1/3},$$

where n is the number density of the ions. The ions are strongly coupled at $\Gamma \geq 1$, if Γ exceeds 172, the ion cloud is crystallized. The temperature of trapped ions is controllable over the range of several milli-Kelvin to 1000 Kelvin or more, where the transition between gas and solid state exists. The deviation from the Boltzmann-Gibbs statistics is expected to appear at the tail component of the velocity distribution function of the strongly-correlated ions, thus stable confinement of ions is required to detect such component. The dynamics of the trapped ions is governed by rf heating, laser cooling (or heating), and Coulomb interaction [3]. Rf heating usually complicates the ion dynamics, therefore rf heating free ion cloud is suitable for our research. One of the advantages of a linear rf trap is the absence of the direct coupling between the ion motion and the rf electric field in the direction of the trap axis. The effect of rf heating in the radial direction is coupled to the axial motion via the Coulomb interaction. However, the radial rf heating is negligible on the trap axis. Therefore, 1-dimensional (1D) ion cloud aligned on the trap axis is the ideal Coulomb system that minimizes the rf heating effect. We report here the performances of our experimental equipment and the preliminary results of generation of ideal 1D ion cloud.

EXPERIMENT

Figure 1 shows the photo of our linear ion trap equipped with a Ca ion source. The assembly is installed in a stainless steel chamber evacuated to less than 3×10^{-10} Torr using an ion getter pump. The linear rf trap is composed of the four cylindrical rf electrodes and the two dc end-electrodes. The rf and dc electrodes are made of copper and stainless steel, respectively. The diagonal distance of the surface of the rf electrodes is 8.8 mm, and the diameter of the rf electrodes is 4 mm. Ions are confined in the radial direction by the oscillating quadrupole field. The voltage applied to the rf electrodes was 180 ~ 210 V at 2.8 MHz. The distance between the two dc electrodes is 20 mm. Confinement of ions in the axial direction is achieved by the static field generated by the dc electrodes. Voltage applied to the dc electrode was 4 ~ 60 V. The calcium vapor oven and the electron gun are arranged face to face on the both sides of the ion trap. To decrease the calcium contamination on the trap electrodes, calcium

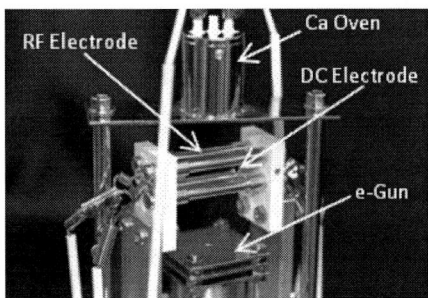

FIGURE 1. Linear rf ion trap equipped with the Ca ion source.

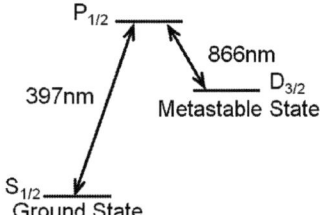

FIGURE 2. Diagram of the relevant states of Ca^+ laser cooling.

vapor is collimated by a skimmer before introducing to the confinement region. Ions generated by electron bombardment are trapped in the ion trap, and cooled using the Doppler laser cooling techniques. The relevant states of $^{40}Ca^+$ laser cooling are shown in fig. 2. All the transitions of the cooling cycle can be driven by commercial laser diodes. Since approximately 90 % of the $P_{1/2}$ state ions decay to the $S_{1/2}$ state, the $S_{1/2}$-$P_{1/2}$ transition, excited by 397-nm laser, is the main cooling-transition. A part of the ions excited on the $P_{1/2}$ state falls into the meatastable $D_{3/2}$ state, therefore an 866-nm laser is used for repumping the ions in the $D_{3/2}$ state to the $P_{1/2}$ state. The frequency of the repumping laser is continuously swept around the resonance frequency to cover the Doppler broadening of the $D_{3/2}$- $P_{1/2}$ transition. Figure 3 shows the schematics diagram of the experimental setup. The 866 and 794-nm laser lights are obtained by extended-cavity diode lasers. The 397 nm laser light is obtained by the second harmonic generation (SHG) of the 794 nm laser light using a nonlinear crystal (LBO) installed in a ring cavity. The spectra of the laser lights are monitored using the Fabry-Perot interferometers to confirm single-mode operation of the ECDLs. Since the temperature of laser-cooled ions is sensitive to the frequency of the 397-nm laser, the short-term fluctuation and the long-term drift of the laser frequency is stabilized using optical feedback from the ring cavity and controlling the resonator length of the ring cavity respectively [4, 5]. The 397 and 866-nm lasers are divided into two components and superimposed each other before introduced from 0° and 45° to the trap axis. The laser induced fluorescence (LIF) signal of the $S_{1/2}$-$P_{1/2}$ transition was used for ion detection. The integration of LIF signal is detected using a photomultiplier tube

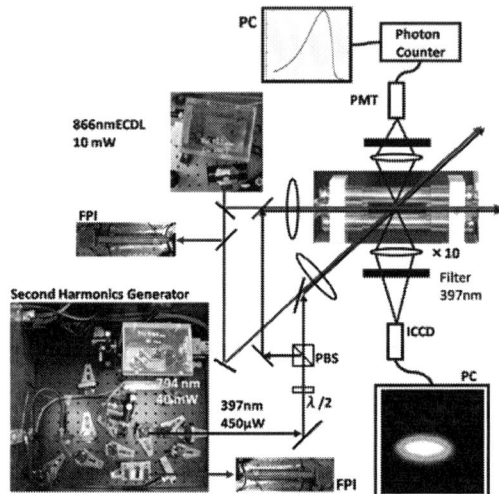

FIGURE 3. Schematics of the laser cooling experiment.

(PMT). The image of ion cloud is observed using an intensified charged coupled device (ICCD) camera with a 10-magnification imaging-lens system.

RESULTS

The stability of the 397-nm light was evaluated by monitoring the position of the fringe of the Fabry-Perot interferometer. The long-term drift of the center frequency was stabilized at 0.01 MHz / min, and the full width at half maximum (FWHM) of the short-term fluctuation was 6 MHz. These values are sufficiently smaller than the natural line width of the $S_{1/2}$-$P_{1/2}$ transition. Figure 4(a) shows the image of a single ion. The halo component caused by the aberration of the imaging system was superimposed on the image of the ion. The FWHM of the ion image and the halo component were 9 μm and 50 μm respectively. The peak intensity of the halo was about 15 % of the ion's peak. The tail component of the halo covered the neighboring ion image, nevertheless the images of the individual ions were still observable (fig. 4(b)).

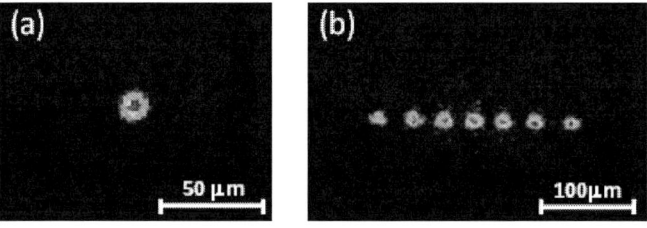

FIGURE 4. Images of (a) a single ion and (b) ion string. Exposure time was 10 s.

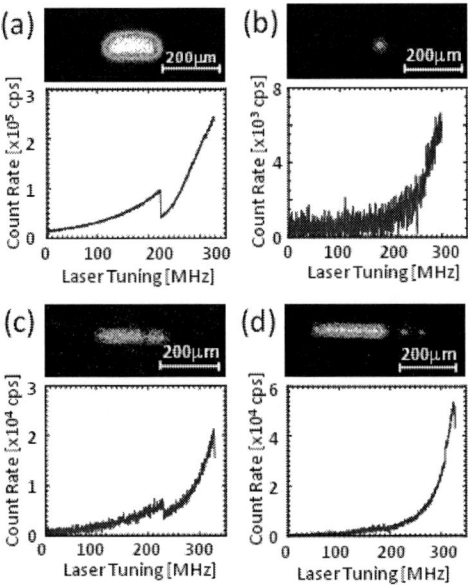

FIGURE 5. Images and LIF spectra of (a) a middle-size ion cloud (V_{RF} = 180 V, V_{DC} = 60 V), (b) a single ion (V_{RF} = 180 V, V_{DC} = 60 V), (c) a string ion cloud confined in deep potential (V_{RF} = 210 V, V_{DC} = 14 V), and (d) a string ion cloud confined in shallow potential (V_{RF} = 210 V, V_{DC} = 4 V).

Figure 5 shows LIF spectra and ICCD images of laser-cooled ion clouds in the linear RF trap. The each picture was taken at the frequency where the maximum LIF signal was observed. The LIF spectra were obtained by sweeping the frequency of the cooling laser from the low frequency side to the resonance center (755.4361 THz). The horizontal axis of the spectrum is the frequency of laser tuning. Here the tuning frequency was calibrated using the fringes of the Fabry-Perot interferometer. The resonance center of the $S_{1/2}$-$P_{1/2}$ transition locates at the vicinity of the frequency where the maximum LIF was observed. However we do not have a frequency standard that is accurate enough to calibrate the absolute value of the laser frequency in the laser cooling experiment, therefore we do not show the resonance center of these spectra. Figure 5(a) shows a typical LIF signal of a middle-size ion cloud. The discrete drop of the LIF signal, which indicates sharp decrease of ion temperature, is the result of the competition between the rf heating and the laser cooling effect [6]. On the other hand, the LIF signal of a single ion did not show a discrete change through the laser-cooling process, since the rf heating effect was negligible (fig. 5(b)). We utilized the difference in the shape of the LIF spectrum to distinguish the dimension of ion clouds. Small ion cloud confined in a deep potential well (fig. 5(c), V_{RF} = 210 V, V_{DC} = 14 V) showed the sudden condensation which was similar to the case of fig. 5(a). On the other hand, the same ion cloud confined in a shallow potential well (fig. 5(d), V_{RF} = 210 V, V_{DC} = 4 V) did not show a clear fluorescence dip. The LIF spectrum of the case fig. 5(d) was similar to the spectrum of a single ion. The both images of the fig. 5(c) and (d) show the 1D crystal as the final state of the laser cooling in the each confinement condition, however, the difference of the LIF spectra indicates the difference of the rf heating effect in the cooling process. It is important to keep the ion cloud in 1D, where the influence of the rf heating is negligible, for the detailed research on the effect of the Coulomb interaction on the statistical characteristics of the ion cloud. However, there are no clear differences in the images observed by the ICCD camera for the different rf heating condition. Therefore, the monitoring of the shape of the LIF spectrum will be a useful indicator for the existence of the rf heating.

CONCLUSION

To study the effect of the long-range correlation among ions on their statistical characteristics, the method of the generation of an ideal ion string was proposed. We have developed the linear rf trap, the cooling diode-lasers, and the LIF observation systems. The SHG laser was stabilized in order to stabilize the state of the ion cloud. The long time drift of the center frequency of the SHG laser was stabilized at 0.01 MHz / min using electrical feedback, and the line width was narrowed to 6 MHz using optical feedback from the external ring cavity. The dimension of ion clouds observed by the ICCD camera is almost the same for gas and liquid states. However, from the shape of the LIF spectrum, the effect of the rf heating was monitored through the laser cooling process.

ACKNOWLEDGMENTS

The authors would like to thank Dr. K. Hayasaka for his valuable advice and discussions. This work was partially supported by the Ministry of Education, Science, Sports and Culture, Grant-in-Aid for Scientific Research (C).

REFERENCES

1. C. Tsallis, *J. Stat. Phys.* **52**, pp. 479 - 487 (1988).
2. For a regularly updated bibliography see http://tsallis.cat.cbpf.br//biblio.htm
3. R. Blümel, C. Kappler, W.Quint, and H.Walther, *Phys. Rev. A* **40**, pp. 808 - 823 (1989).
4. K. Hayasaka, *Opt. Commun.* **206**, pp. 401-409 (2002).
5. K. Matsubara, S. Uetake, H. Ito, Y. Li, K. Hayasaka, and M. Hosokawa, *Jpn. J. Appl. Phys.* **44**, pp. 229 - 230 (2005).
6. L. Hornekær and M. Drewsen, *Phys. Rev. A* **66**, 013412 (2002).

Barium Ions for Quantum Computation

M. R. Dietrich, A. Avril, R. Bowler, N. Kurz, J. S. Salacka, G. Shu and B. B. Blinov

University of Washington Department of Physics, Seattle, Washington, 98195

Abstract. Individually trapped $^{137}Ba^+$ in an RF Paul trap is proposed as a qubit candidate, and its various benefits are compared to other ionic qubits. We report the current experimental status of using this ion for quantum computation. Future plans and prospects are discussed.

Keywords: Quantum Computation, Barium, Trapped Ions
PACS: 03.67.Lx, 37.10.Ty, 42.50.Dv

INTRODUCTION

At present, trapped ions are the leading contender for use as a qubit in quantum computation schemes. This is a result of the high degree of motional control possible over the ion, the availability of long established techniques for quantum manipulation of trapped ions, and the success of shelving schemes as a highly efficient readout mechanism. To date, Be^+ [1], Ca^+ [2, 3], Cd^+ [4], Mg^+ [5], Sr^+ [6], and Yb^+ [7] have all been demonstrated as possible ionic qubits. It is possible to create ionic qubits using either two hyperfine levels [8] or, in some species, two levels separated by an optical transition [2] as the computational basis. We propose to use the ground state hyperfine levels of $^{137}Ba^+$ as a qubit. Although Ba^+ was the first ion trapped in isolation [9], $^{137}Ba^+$ was not trapped for another 20 years [10]. The spectroscopic properties of Ba^+ have been carefully studied since then because of its potential applications as an optical frequency standard [11] and in a test of parity non-conservation [12]. It has several desirable qubit properties, including visible wavelength transitions, high natural abundance of the ^{137}Ba isotope, and a long lived shelving state. Here we demonstrate single qubit initialization, rotations and readout on this new qubit, and discuss the future directions of our work.

EXPERIMENTAL SETUP

Ba^+, like some of its qubit competitors such as Ca^+ and Sr^+, has an energy level structure that includes two low lying, long lived D states - see Fig. 1(a). In contrast with these ions, however, all of the Ba^+ dipole transitions lie in the visible spectrum, which greatly simplifies laser alignment. The doppler cooling consists of a blue 493 nm transition from the ground state to the $P_{1/2}$ state, which has a branching ratio of 0.244 [13] to $D_{3/2}$. Because this state is long lived, a repump laser at 650 nm is necessary for continuous cooling. However, the upper D state, $D_{5/2}$, is isolated from the cooling cycle and so constitutes a "dark" state which can be used for high fidelity readout. Its lifetime of 35 s helps reduce the overall error rate during readout, compared to the

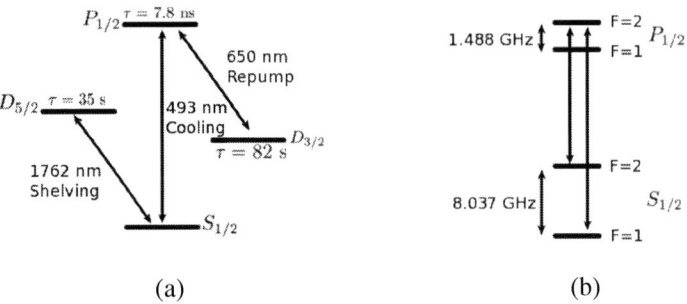

FIGURE 1. Energy level diagram for Ba$^+$. (a) The basic cooling and shelving scheme. (b) A detail of the 493 cooling transition, showing the relevant transitions. The (F=2,m_F=0) to (F=2,m_F=0) π transition is actually forbidden, and so an elliptical polarization must be used for cooling.

relatively short lifetime of, for example, Ca$^+$ at 1 s. The predominant isotope of Ba is 138 (72% abundance), which has no nuclear spin and so no hyperfine structure to use as a qubit. When trapping the odd isotope of Ba$^+$ (11% abundance), all the various hyperfine levels must be addressed by introducing sidebands onto the two cooling lasers. For the red 650 nm laser, the $D_{3/2}$ levels lie close enough [14] that not all possible transitions need to be covered and so only three modulation frequencies are used - 614, 539, and 394 MHz. For the blue, modulation is introduced with frequency equal to the ground state hyperfine splitting, and the carrier is set such that all ground states are excited only into the $P_{1/2}$ (F=2) manifold, see Fig. 1(b). This is advantageous not only for the optical pumping reasons discussed below, but also because one avoids the $S_{1/2}$ (F=1) to $P_{1/2}$ (F=1) transition which is extremely weak as a result of small, destructively interfering geometric factors [15].

The ion can be excited directly from the F=2 ground state to the $D_{5/2}$ level using a 1762 nm fiber laser (Koheras Adjustic). While in this shelved state, the ion will not flouresce when illuminated with the cooling lasers. A bright ion can be distinguished from a dark one with nearly perfect fidelity after a couple ms of observation time, resulting in highly reliable readout. Because the transition is so weak (E2 transition), resonant excitation would require a carefully stabilized laser. However, using adiabatic passage, it is possible to perform highly efficient population transfers using only a poorly stabilized laser [16]. Using the 10 mW of available power, simulation indicates that this should be possible on the ms time scale. Because of the many magnetic sublevels of the $D_{5/2}$ state, however, it is necessary to perform some stabilization to prevent accidental excitation into an undesired state.

The ground state of ^{137}Ba$^+$ is split due to the hyperfine interaction by about 8.037 GHz [17], which makes this isotope usable as a hyperfine qubit. We optically pump into the upper m_F=0 state with π polarized 493 nm light, since parity symmetry forbids the $S_{1/2}$ (F=2,m_F=0) to $P_{1/2}$ (F=2,m_F=0) transition. After state preparation, we can cause direct Rabi flops between the upper and lower magnetically insensitive m_F=0 states using microwaves at the known hyperfine frequency, exposed for varying periods of time.

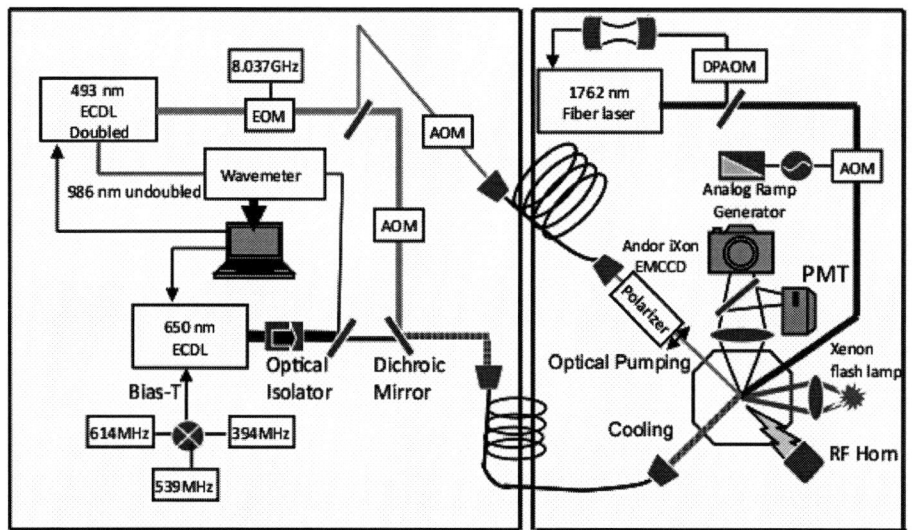

FIGURE 2. A schematic of the experimental setup. The cooling lasers are locked using a commercial wavemeter, which feeds back to the laser systems. Acousto-optic modulators (AOMs) are used for high speed shuttering and, in the case of the 1762, frequency modulation and shifting (Double Pass AOM). Barium ionization is achieved through photoionization with a Xe flash lamp. A mirror on a motorized mount changes the detection from a CCD camera to a PMT.

The cooling light is provided by two external cavity diode lasers (ECDL), one at 650 nm (Toptica DL-100) and the other at 493 nm (Toptica SHG-110), the latter of which is a 986 nm diode frequency doubled in a bow-tie enhancement cavity. The available power is 10 mW and 20 mW respectively. Both lasers are stabilized to within about 3 MHz using a high-precision wavelength meter (HighFinesse WS-7). The blue is then modulated using a resonant EOM at 8.037 GHz (New Focus model 4851), while the red has its drive current modulated directly with a bias-T. The two are then combined using a dichroic mirror and coupled into a single mode fiber, which provides colinearization and mode cleaning. The fiber also allows us to send the cooling laser light into another room where the active ion trap presently resides. A small fraction, about 10 μW, of 493 nm light is split off after the EOM for the optical pumping. The 1762 nm fiber laser is output onto the second table, where it is stabilized using a high finesse Zerodur cavity with 500 MHz free spectral range suspended in a vacuum chamber which is temperature stabilized to within about 10 mK. The transmitted laser intensity is monitored to maintain the lock. The adiabatic passage is achieved by driving an AOM with a linear analog ramp, ensuring maximum adiabaticity. To load Ba into the trap, we heat a sample of metallic barium contained in an alumina cylinder to several hundred degree Celsius to create an atomic beam and then use a Xe flash lamp to photoionize. The abundance of ^{137}Ba is sufficiently high that one or two ions can be reliably trapped without the need for isotope selective photoionization.

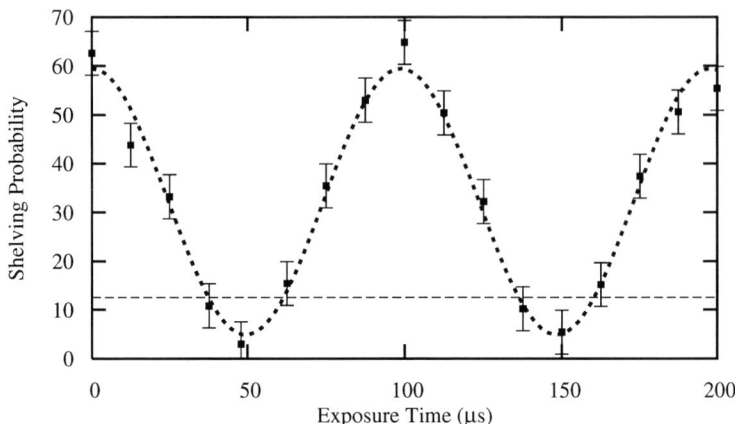

FIGURE 3. Probability of shelving as a function of microwave exposure time. The Rabi frequency here is 10.1 kHz. The maximum shelving probability is limited by the stability of the shelving laser, and the minimum by optical pumping efficiency and microwave frequency detuning. The minimum shelving observed was 4.8%, and the contrast is 54%. $\chi_\nu^2 = .55$ for this fit, with $\nu = 13$. The error bars are statistical. The dashed line indicates the 12.5% minimum which would be observed in the absence of optical pumping.

CURRENT STATUS

After qubit state preparation by optical pumping we drive Rabi flops by applying microwaves and detect the final state with the 1762 nm laser as described above. The resulting detected Rabi flops are shown in Fig. 3. At the time of publication, the 1762 nm laser stabilization was not complete, so that the 1762 frequency sweep crossed several transitions, resulting in only 60% efficient population transfer from the $S_{1/2}$ (F=2, m_F=0) to the $D_{5/2}$ (F=3) and (F=4) manifold. It should be noted that without optical pumping the lowest possible shelving probability would be 12.5% since all ground state m_F levels would be populated evenly, and so this data simultaneously illustrates optical pumping.

We have also recently demonstrated a full optical Rabi flop on a single ^{138}Ba$^+$ using a single ultrafast laser pulse of 400 fs duration on the $S_{1/2}$ to $P_{3/2}$ transition at 455 nm [18]. This allowed the measurement of the branching ratios of the $P_{3/2}$ state with high precision. Branching ratio measurements provide an experimental test of computational models of Ba$^+$ atomic structure, which are important for the parity non-conservation test [12]. Also, when compared against astronomical measurements of the branching ratio, it provides a bound on prehistoric variations in α. The ultrafast excitation of single Ba ions paves the way to the ion-photon [19] and remote ion-ion [20] entanglement.

FUTURE PLANS

If tuned to the $P_{1/2}$ transition, ultrafast Rabi flops could be made to apply a state dependent impulse on the ion, which would allow us to perform ultrafast gates, such as the García-Ripoll phase gate [21], on two ions in the same trap. The benefits of such a gate are its intrinsic speed and that it does not necessitate cooling to the ground state of motion. This, combined with a laser for stimulated Raman transitions between the hyperfine levels, will form the basis of quantum computing with barium.

A more immediate objective with Ba^+ is the remote entanglement of two ions seperated by a kilometer or more. Remote entanglement of ions is accomplished by first exciting each ion with the ultrafast laser and allowing it to spontaneously decay, resulting in a photon whose frequency or polarization state is entangled with the final spin state of the relaxed ion. Once this is done with two separate ions, those photons can be jointly measured in the appropriate parity basis, resulting in an entangled state between the two ions. Entanglement between a photon and ion was originally seen in Cd^+ [19] and between two ions just recently in Yb^+ [20]. At present, the distance of entanglement is partly limited by the length that short wavelengths of light travel through an optical fiber, another potential advantage of Ba^+, since its cooling wavelength is longer than these other examples. The ability to remotely entangle qubits has application not only for quantum repeaters [22, 23] but also for loophole free Bell inequality tests [24]. Necessary to the success of such an experiment is the ability to perform every operation very quickly, on the μs time scale. In order to achieve such time scales, the power of the 1762 nm laser will have to be amplified, and the collection efficiency of the ion florescence will have to be increased. The former task can be accomplished using a Tm based fiber amplifier presently under development. The latter task will involve a high numerical aperture light collecting mirror placed inside the vacuum, near the trap itself. This will greatly increase the light collection solid angle, and thereby decrease the bright/dark discrimination time, which is currently limited to a ms. This modified trap design is presently in testing.

ACKNOWLEDGEMENTS

We would like to give special thanks to Sanghoon Chong, Tom Chartrand, Adam Kleczewski, Viki Mirgon, Joseph Pirtle and Edan Shahar for their various contributions to the experiment. This work was supported by NSF AMO program, the ARO DURIP grant, and the University of Washington Royalty Research Fund.

REFERENCES

1. Q. A. Turchette, C. S. Wood, B. E. King, C. J. Myatt, D. Leibfried, W. M. Itano, C. Monroe, and D. J. Wineland, Phys. Rev. Lett. **81**, 3631 (1998).
2. R. Blatt, H. Haffner, C. F. Roos, C. Becher, and F. Schmidt-Kaler, Quant. Inf. Proc. **3**, 61 (2004).
3. D. M. Lucas, B. C. Keitch, J. P. Home, G. Imreh, M. J. McDonnell, D. N. Stacey, D. J. Szwer, and A. M. Steane (2007), quant-ph/0710.4421.
4. P. J. Lee, B. B. Blinov, K. Brickman, L. Deslauriers, M. J. Madsen, R. Miller, D. L. Moehring, D. Stick, and C. Monroe, Opt. Lett. **28**, 1582 (2003).
5. T. Schaetz, A. Friedenauer, H. Schmitz, L. Petersen, and S. Kahra, J. Mod. Optics **54**, 2317 (2007).

6. V. Letchumanan, G. Wilpers, M. Brownnutt, P. Gill, and A. G. Sinclair, Phys. Rev. A **75**, 063425 (2007).
7. S. Olmschenk, K. C. Younge, D. L. Moehring, D. Matsukevich, P. Maunz, and C. Monroe, Phys. Rev. A **76**, 052314 (2007).
8. B. B. Blinov, D. Leibfried, C. Monroe, and D. J. Wineland, Quant. Inf. Proc. **3**, 45 (2004).
9. W. Neuhauser, M. Hohenstatt, P. E. Toschek, and H. Dehmelt, Phys. Rev. A. **22**, 1137 (1980).
10. R. G. DeVoe and C. Kurtsiefer, Phys. Rev. A **65**, 063407 (2002).
11. T. W. Koerber, M. H. Schacht, K. R. G. Hendrickson, W. Nagourney, and E. N. Fortson, Phys. Rev. Lett. **88**, 143002 (2002).
12. N. Fortson, Phys. Rev. Lett. **70**, 2383 (1993).
13. M. D. Davidson, L. C. Snoek, H. Volten, and A. Dönszelmann, Astron. Astrophys. **255**, 457 (1992).
14. R. E. Silverans, G. Borghs, P. De Bisschop, and M. Van Hove, Phys. Rev. A **33**, 2117 (1986).
15. H. J. Metcalf and P. van der Straten, *Laser Cooling and Trapping* (Springer, 2001), ISBN 978-0-387-98728-6.
16. C. Wunderlich, T. Hannemann, T. Körber, H. Häffner, C. Roos, W. Hänsel, R. Blatt, and F. Schmidt-Kaler, J. Mod. Optics **54**, 1541 (2007), `quant-ph/0508159`.
17. R. Blatt and G. Werth, Phys. Rev. A **25**, 1476 (1982).
18. N. Kurz, M. R. Dietrich, G. Shu, R. Bowler, J. Salacka, V. Mirgon, and B. B. Blinov, Phys. Rev. A **77**, 060501 (2008).
19. B. B. Blinov, D. L. Moehring, L.-M. Duan, and C. Monroe, Nature **428**, 153 (2004).
20. D. L. Moehring, P. Maunz, S. Olmschenk, K. C. Younge, D. N. Matsukevich, L. M. Duan, and C. Monroe, Nature **449**, 68 (2007).
21. J. J. García-Ripoll, P. Zoller, and J. I. Cirac, Phys. Rev. Lett. **91**, 157901 (2003).
22. H.-J. Briegel, W. Dür, J. I. Cirac, and P. Zoller, Phys. Rev. Lett. **81**, 5932 (1998).
23. L. M. Duan, B. B. Blinov, D. L. Moehring, and C. Monroe, Quant. Inf. and Comp. **4**, 165 (2004), `quant-ph/0401020`.
24. C. Simon and W. T. M. Irvine, Phys. Rev. Lett. **91**, 110405 (2003).

Quantum Optical Heating in Sonoluminescence Experiments

Andreas Kurcz, Antonio Capolupo and Almut Beige

School of Physics and Astronomy, University of Leeds, Leeds, LS2 9JT, United Kingdom

Abstract. Sonoluminescence occurs when tiny bubbles filled with noble gas atoms are driven by a sound wave. Each cycle of the driving field is accompanied by a collapse phase in which the bubble radius decreases rapidly until a short but very strong light flash is emitted. The spectrum of the light corresponds to very high temperatures and hints at the presence of a hot plasma core. While everyone accepts that the effect is real, the main energy focussing mechanism is highly controversial. Here we suggest that the heating of the bubble might be due to a weak but highly inhomogeneous electric field as it occurs during rapid bubble deformations [A. Kurcz *et al.* (submitted)]. It is shown that such a field couples the quantised motion of the atoms to their electronic states, thereby resulting in very high heating rates.

Keywords: Sonoluminescence, Ion Trapping.
PACS: 78.60.Mq, 43.25.+y, 37.10.Ty

INTRODUCTION

Sonoluminescence is a phenomenon that derives from the acoustic cavitation of noble gas atoms [1]. There are two classes of sonoluminescence: multi-bubble [2, 3] and single-bubble sonoluminescence [4, 5]. Single bubble sonoluminescence is characterized by the emission of a strong light flash over a very short period of time from a single extremely hot gas bubble. Under appropriate conditions, the acoustic force on a bubble can balance against its buoyancy, holding a bubble stable in the liquid by acoustic levitation. Such a bubble is typically quite small compared to an acoustic wavelength and is capable to confine the particles of the trapped van der Waals gas close to their covolume. For specialized conditions, a single, stable, oscillating gas bubble can be forced into such large amplitude pulsations that it produces sonoluminescence during each and every acoustic cycle.

A typical single-bubble sonoluminescence cycle is shown in Fig. 1(a). Most of the cycle, the bubble behaves isothermal (c.f. Fig. 1(b)). Point A marks the beginning of the collapse phase in which the bubble approaches its minimum radius of about $0.5\,\mu$m very rapidly with supersonic speed. Here, the bubble becomes thermally isolated from the surrounding liquid. In Point B, the temperature within the bubble significantly increases with a heating rate of $10^{10} - 10^{11}$ K/s and a strong light flash emerges which last for about 40 ps. Point C denotes the beginning of the expansion phase in which the bubble oscillates around its equilibrium radius until it regains stability.

The emitted light mainly consist of a continuum of blackbody or Bremsstrahlung radiation. Detailed measurements of the light spectra indicate temperatures above 10^4 K [6, 7, 8]. It is even possible to observe light emission in the ultraviolet regime which hints

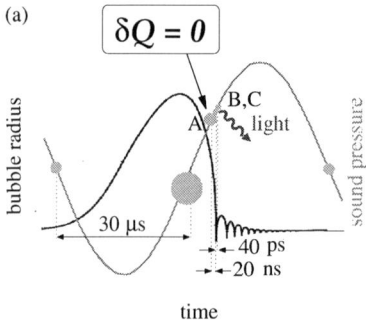 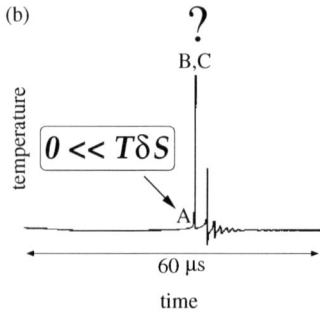

FIGURE 1. A typical single-bubble sonoluminescence cycle. (a): Time dependence of the driving sound pressure and the corresponding bubble radius. (b): Time dependence of the temperature.

at temperatures of about 10^6 K in a bubble driven at 1 Mhz [9]. Emission lines from transitions between high energy states of noble gas atoms which cannot be populated thermally [10, 11] point at the formation of an *opaque* plasma core [13]. Evidence for a plasma core has also been found in multi-bubble sonoluminescence experiments [12].

The time dependence of the bubble radius and its nearly adiabatic compression are theoretically well understood up to a certain point when it approaches the minimum radius [5, 13]. What is the state of the bubble during the last part of the collapse phase and the conditions that lead to these enormous heating rates up to very high temperatures is still controversial. Here we summarise an idea which suggests that the heating is due to the presence of a highly inhomogenous electric field as it occurs during rapid bubble deformations [14]. This field couples the motion of the noble atoms to their electronic degrees of freedom. When combined with spontaneous emission from the atoms, a quantum optical heating process can occur. Similar couplings are responsible for the cooling of ions in ion trap experiments [15].

THE BASIC IDEA

Approaching its minimum radius close to point B in Fig. 1(a), the bubble is no longer in a thermal equilibrium. Suddenly, an increase in entropy occurs which is based on highly irreversible processes. It causes a temperature increase much higher than what can be caused by thermodynamic heating processes (c.f. Fig. 1(b)). In the following we address two questions: Why does the bubbles need to be filled with noble gas atoms? What is the main energy focussing mechanism during the collapse phase of the bubble?

Close to point B, the mean distance between the noble gas atoms becomes so small that interactions between them can be described by a Lennard-Jones potential. Indeed, the physical condition of the bubble becomes that of a solid state system. The atoms experience an equilibrium between repulsive interatomic forces due to overlapping orbitals and attractive forces due to the van der Waals interaction. Thus, any significant gain in temperature has to be caused by vibrational motion driven into the quantum regime. Furthermore, the presence of light requires the assumption of an *open* quantum system.

To model the resulting strong confinement of the atoms, we place each of them into an approximately harmonic trapping potential. This allows us to quantise the atomic motion during the collapse phase, just before the maximum compression of the bubble. Around this point, the motional states of each atom can be described by phonons with frequency v. In the next section, we show that the gradient of an electric field inside the bubble establishes a coupling between the electronic and the quantised motional states of each noble gas atom. The origin of the field can be explained by an inhomogeneous charge distribution of ionized species from the dissolved liquid due to rapid bubble deformations.

For simplicity we assume that the atoms are effective two-level systems with ground state $|0\rangle$ and excited state $|1\rangle$. The corresponding interaction Hamiltonian contains terms that result in the excitation and de-excitation of each atom accompanied by the creation and the annihilation of a phonon. Also crucial is the presence of a large spontaneous decay rate Γ of the excited state $|1\rangle$ which keeps the atoms predominantly in their ground state. Although these processes are highly non-resonant, they result in a significant change of the mean phonon number per atom and increase the temperature inside the bubble by many orders of magnitude, even within a few nanoseconds.

Suppose an atom is initially in its ground state and possesses exactly m phonons, as shown in Fig. 2a. We denote this state by $|0,m\rangle$. Notice that phonons are bosons which are described by annihilation operators b with $[b,b^\dagger] = 1$. Consequently, a transition into the state $|1,m+1\rangle$ occurs with a rate proportional to $\sqrt{m+1}$, while the rate for a transition into the state $|1,m-1\rangle$ scales only as $\sqrt{m}$. Since the spontaneous decay rate of the atom is relatively large, such a transition is immediately followed by an irreversible and predominantly non-radiative transition back into $|0\rangle$. This transfers the atom either into its initial state $|0,m\rangle$ or into the states $|0,m-1\rangle$ and $|0,m+1\rangle$, respectively. The net effect is an increase of the mean phonon number per atom, i.e. heating, since the phonon population in the latter state is higher than the phonon population in $|0,m-1\rangle$.

THE TIME EVOLUTION OF THE SYSTEM

We now consider a *single* noble gas atom at the position $\mathbf{r}$. This atom is typical for the many atoms inside the bubble. Its dipole Hamiltonian equals

$$H_{\text{int}} = e\mathbf{D} \cdot \mathbf{E}(\mathbf{r}) \tag{1}$$

with the (real) atomic dipole moment

$$\mathbf{D} = \mathbf{D}_{01}\,\sigma^- + \text{H.c.}, \tag{2}$$

$\sigma^+ \equiv |1\rangle\langle 0|$, $\sigma^- \equiv |0\rangle\langle 1|$, and where $\mathbf{E}$ is the electric field inside the bubble. For simplicity, we assume that all field components point in the direction of a single unit vector $\hat{\mathbf{k}}$. This allows us to write $\mathbf{E}(\mathbf{r})$ as

$$\mathbf{E}(\mathbf{r}) = \sum_k \mathbf{E}_k\, e^{ik\hat{\mathbf{k}}\cdot\mathbf{r}} + \text{c.c.} \tag{3}$$

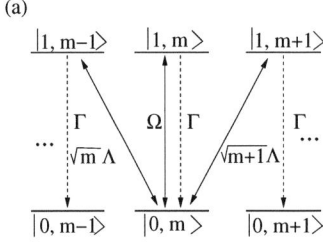 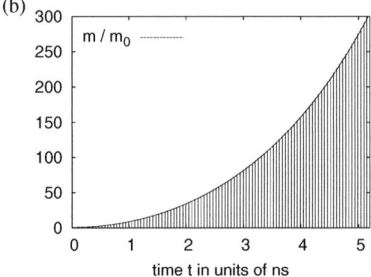

FIGURE 2. (a): Level configuration of a single atom-phonon system indicating the immediately relevant transitions, if the atom is initially in $|0,m\rangle$. Ω and Λ denote coupling constants and Γ is the spontaneous decay rate of level 1. (b): The mean phonon number m as a function of time for $\nu = 10\,\mathrm{MHz}$ while $\Omega = 10^6\,\mathrm{Hz}$, $\Lambda = 10^{12}\,\mathrm{Hz}$, $\Gamma = 10^{13}\,\mathrm{Hz}^{-1}$, and $\omega_0 = 10^{15}\,\mathrm{Hz}$. Good agreement is found between the numerical solution of the full rate equations (9) and (10) and Eq. (13) (shaded area).

with amplitudes $\mathbf{E}_k$ and wave vectors $\mathbf{k} = k\hat{\mathbf{k}}$. Moreover, we consider the atomic motion in the $\hat{\mathbf{k}}$-direction as quantised with b being the corresponding phonon annihilation operator. Then $\hat{\mathbf{k}} \cdot (\mathbf{r} - \mathbf{R}) = \Delta x (b + b^\dagger)$. Here $\mathbf{R}$ is the current equilibrium position of the noble gas atom with mass M and $\Delta x = \sqrt{\hbar/2M\nu}$ is the width of its ground state wave function in the respective vibrational mode. If the atom is well localized within the wavelength of its trapping potential, the Lamb-Dicke approximation allows us to assume that $\exp(ik\hat{\mathbf{k}} \cdot (\mathbf{r} - \mathbf{R})) = 1 + ik\Delta x (b + b^\dagger)$ [15, 16]. Substituting this into Eq. (1), we obtain the interaction Hamiltonian

$$H_{\text{int}} = \hbar\Omega(\sigma^- + \sigma^+) + \hbar\Lambda(b + b^\dagger)(\sigma^- + \sigma^+) \qquad (4)$$

with the (real and positive) coupling constants

$$\begin{aligned}
\Omega &\equiv (2e/\hbar)\sum_k \mathbf{D}_{01} \cdot \mathrm{Re}\left(\mathbf{E}_k e^{ik\hat{\mathbf{k}}\cdot\mathbf{R}}\right), \\
\Lambda &\equiv -(2e\Delta x/\hbar)\sum_k k\mathbf{D}_{01} \cdot \mathrm{Im}\left(\mathbf{E}_k e^{ik\hat{\mathbf{k}}\cdot\mathbf{R}}\right).
\end{aligned} \qquad (5)$$

This Hamiltonian is essentially a Jaynes-Cummings Hamiltonian with Λ being proportional to the gradient of Ω in the direction of the quantised motion of the atom, i.e. $\Lambda = \Delta x \hat{\mathbf{k}} \cdot \nabla_{\mathbf{R}}\Omega$. A strong atom-phonon coupling therefore does not necessarily require the presence of a strong electric field. It only requires a highly inhomogeneous field inside the bubble. In the following, we neglect interactions between the noble gas atoms other than the ones already included in the harmonic trapping potential of each particle. Dissipation in form of spontaneous photon emission from the atomic state $|1\rangle$ is taken into account by the master equation [16]

$$\dot{\rho} = -\frac{i}{\hbar}\left[H_{\text{int}} + \hbar\omega_0\sigma^+\sigma^- + \hbar\nu b^\dagger b, \rho\right] + \Gamma\left[\sigma^-\rho\sigma^+ - \frac{1}{2}\sigma^+\sigma^-\rho - \frac{1}{2}\rho\sigma^+\sigma^-\right]. \qquad (6)$$

Here $\hbar\omega_0$ and $\hbar\nu$ are the energy of the atomic state $|1\rangle$ and of a single phonon.

Eq. (6) can now be used to obtain a closed set of rate equations. Its major quantities are the phonon number $m \equiv \langle b^\dagger b \rangle$ and

$$X_{1,2} \equiv \langle \sigma_{1,2} \rangle, \quad X_3 \equiv \langle \sigma^+\sigma^- - \sigma^-\sigma^+ \rangle, \quad Y_1 \equiv \langle b+b^\dagger \rangle, \quad Y_2 \equiv i\langle b-b^\dagger \rangle,$$
$$Y_3 \equiv \langle b^2 + b^{\dagger 2} \rangle, \quad Y_4 \equiv i\langle b^2 - b^{\dagger 2} \rangle, \quad Z_{1,2} \equiv \langle \sigma_{1,2}(b+b^\dagger) \rangle, \quad Z_{3,4} \equiv i\langle \sigma_{1,2}(b-b^\dagger) \rangle \quad (7)$$

with the Pauli operators $\sigma_1 \equiv \sigma^+ + \sigma^-$ and $\sigma_2 \equiv i(\sigma^- - \sigma^+)$. Here we assume

$$\omega_0 \gg \nu, \Gamma, \Omega, \Lambda \text{ and } m \gg 1 \qquad (8)$$

and approximate the expectation value of operators of the form $\langle B\sigma_3 \rangle$ by $\langle B \rangle \langle \sigma_3 \rangle$. The latter applies when the expectation value of B is about the same for an atom in $|0\rangle$ and for an atom in $|1\rangle$. Eq. (6) then yields

$$\dot{m} = \Lambda Z_3, \quad \dot{X}_3 = 2(\Omega X_2 + \Lambda Z_2) - \Gamma(X_3+1), \quad \dot{Y}_1 = -\nu Y_2,$$
$$\dot{Y}_2 = 2\Lambda X_1 + \nu Y_1, \quad \dot{Y}_3 = -2(\nu Y_4 + \Lambda Z_3), \quad \dot{Y}_4 = 2(\nu Y_3 + \Lambda Z_1), \qquad (9)$$

and

$$\dot{X}_1 = -\omega_0 X_2, \quad \dot{X}_2 = -2(\Omega + \Lambda Y_1)X_3 + \omega_0 X_1, \quad \dot{Z}_1 = -\omega_0 Z_2, \quad \dot{Z}_3 = 2\Lambda - \omega_0 Z_4,$$
$$\dot{Z}_2 = -2(\Omega Y_1 + \Lambda Y_3 + 2\Lambda m)X_3 + \omega_0 Z_1, \quad \dot{Z}_4 = -2(\Omega Y_2 + \Lambda Y_4)X_3 + \omega_0 Z_3 \qquad (10)$$

up to first order in $1/\omega_0$. In the beginning of each sonoluminescence cycle the particles experience neither a strong trapping potential nor the presence of an inhomogeneous electric field inside the bubble. We can therefore assume that the coherences defined in Eq. (7) are initially zero and that the atom is in its ground state. Condition (8) allows us to simplify the above rate equations via an adiabatic elimination of Eq. (10). Doing so we obtain a set of equations where the derivatives of X_3, Y_1, and Y_2 decouple from the rest. Solving them for the case of a relatively strong atom-phonon coupling constant Λ with $\Lambda \gg \Omega$ and $4\Lambda^2 > \nu\omega_0$ yields

$$X_3(t) = -1, \quad Y_1(t) = \frac{4\nu\Omega\Lambda}{\lambda^2 \omega_0} \cdot [\cosh(\lambda t) - 1], \quad Y_2(t) = -\frac{4\Omega\Lambda}{\lambda \omega_0} \cdot \sinh(\lambda t) \qquad (11)$$

with $\lambda \equiv \nu(4\Lambda^2/\nu\omega_0 - 1)^{1/2}$ up to first order in $1/\omega_0$. For times t of the order of $1/\lambda$, Y_1 and Y_2 are of the order of $1/\omega_0$. Taking this into account, we find that $Z_1 = -2\Lambda(2m+Y_3)/\omega_0$ and $Z_3 = -2\Lambda Y_4/\omega_0$ in first order in $1/\omega_0$. The variables m, Y_3, and Y_4 in Eq. (9) hence evolve according to

$$\dot{m} = -\frac{2\Lambda^2}{\omega_0} Y_4, \quad \dot{Y}_3 = \frac{2(2\Lambda^2 - \nu\omega_0)}{\omega_0} Y_4, \quad \dot{Y}_4 = -\frac{8\Lambda^2}{\omega_0} m - \frac{2(2\Lambda^2 - \nu\omega_0)}{\omega_0} Y_3. \qquad (12)$$

For $m(0) = m_0$ and $Y_3(0) = Y_4(0) = 0$, this yields

$$m(t) = m_0 + \frac{8\Lambda^4}{\lambda^2 \omega_0^2} \cdot m_0 \sinh^2(\lambda t). \qquad (13)$$

As one can see in Fig. 2(b), Eq. (13) describes an approximately exponential heating process as long as a relatively large decay Γ secures that the atom remains in the ground state predominantly [1]. Taking into account typical experimental parameters, the phonon energy in the bubble can easily increase by a factor ten or more, even within a few nanoseconds. Using the relation $m \cdot \hbar \nu = k_B T$, our model can easily predict temperatures well above 10^4 K inside the bubble.

CONCLUSION

We attribute the sudden concentration of energy in sonoluminescence experiments to the heating of strongly confined noble gas atoms by a highly inhomogeneous electric field. The time evolution of each atom is dominated by non-energy conserving processes, which result in a permanent increase of its mean phonon number m when combined with spontaneous emission. Our model does not contradict current models for the description of sonoluminescence experiments, but explains previously controversial aspects of this phenomenon. It is based on a quantum optical approach that is routinely used to describe the laser cooling of tightly trapped ions [15].

ACKNOWLEDGMENTS

A. B. acknowledges a James Ellis University Research Fellowship from the Royal Society and the GCHQ. This work was moreover supported in part by the EU Research and Training Network EMALI and the UK Research Council EPSRC.

REFERENCES

1. H. Frenzel and H. Schultes, *Z. Phys. Chem. Abt. B* **27B**, 421–424 (1934).
2. A. J. Walton and G. T. Reynolds, *Adv. Phys.* **33**, 595–600 (1984).
3. W. B. McNamara III, Y. Didenko, and K. S. Suslick, *Nature* **401**, 772–775 (1999).
4. D. F. Gaitan, L. A. Crum, C. C. Church, and R. A. Roy.,*J. Acoust. Soc. Am.* **91**, 3166–3183 (1992).
5. M. P. Brenner, S. Hilgenfeldt, and D. Lohse, *Rev. Mod. Phys.* **74**, 425–484 (2002).
6. B. P. Barber and S. J. Putterman, *Nature* **352**, 318 (1991).
7. R. A. Hiller and S. J. Putterman, *Phys. Rev. Lett.* **69**, 1182 (1992).
8. G. Vazquez, C. Camara, S. J. Putterman, and K. Weninger, *Opt. Lett.* **26**, 575 (2001).
9. C. Camara, S. J. Putterman, and E. Kirilov, *Phys. Rev. Lett.* **92**, 124301-1 (2004).
10. D. J. Flannigan and K. S. Suslick, *Nature* **434**, 52 (2005).
11. D. J. Flannigan and K. S. Suslick, *Phys. Rev. Lett.* **99**, 134301 (2007).
12. N. C. Eddingsaas and K. S. Suslick, *J. Am. Chem. Soc.* **129**, 3838–3839 (2007).
13. K. S. Suslick and D. J. Flannigan, *Annu. Rev. Phys. Chem.* **59**, 659–683 (2008).
14. A. Kurcz, A. Capolupo, and A. Beige (submitted).
15. D. Leibfried, R. Blatt, C. Monroe, and D. Wineland, *Rev. Mod. Phys.* **75**, 281–324 (2003).
16. C. C. Gerry and P. L. Knight, *Introductory Quantum Optics*, Cambridge University Press, Cambridge, 2005.

---

[1] The assumption of a relatively high spontaneous decay rate can be justified by the presence of collective effects inside the van der Waals gas formed by the noble gas atoms.

SECTION II

TOROIDAL TRAPS

Achieving Long Confinement in a Toroidal Electron Plasma

J.P. Marler, J. Smoniewski, Bao Ha and M. R. Stoneking

Lawrence University, Appleton, WI, 54911

Abstract.
We observe the m=1 diocotron mode in a partial toroidal trap, and use it as the primary diagnostic for observing the plasma confinement. The frequency of the m=1 mode, which is approximately proportional to the trapped charge, decays on a three second timescale. The confinement time exceeds, by at least an order of magnitude, the confinement observed in all other toroidal traps for non-neutral plasmas and approaches the theoretical limit set by magnetic pumping transport. Numerical simulations that include toroidal effects are employed to accurately extract plasma charge, equilibrium position and m=1 mode amplitude from the experimental data. Future work will include attempts to withdraw the electron source in order to study confinement in a full torus.

Keywords: electron plasmas, non-neutral, toroidal confinement
PACS: 52.27.Jt, 52.55.Hc, 52.55.Dy

INTRODUCTION

Issues of equilibrium, stability, dynamics and limitations on confinement for toroidal non-neutral plasmas have been addressed in two experimental devices at Lawrence University. In the Lawrence Non-neutral Torus (LNT), electron plasmas were trapped in a 300° toroidal arc by application of gate potentials to grids in a modest, 200 G, magnetic field[1]. The existence of a dynamical equilibrium was demonstrated in that device by achieving confinement times that were long compared to all of the single-particle drift timescales. The longest confinement times (18 ms) were reached by employing active feedback to suppress growth of an unstable $m = 1$ diocotron mode[2]. The desire to study the dynamics of electron plasmas under nearly steady state conditions in a full torus, and observe toroidal effects on confinement motivated the design and construction of a new device. The design for the Lawrence Non-neutral Torus II (LNT II) incorporated the ability to generate a ~1 kG magnetic field in high vacuum conditions, with a fully sectored conducting boundary, and a means of making the transition from trapping in a toroidal sector to a fully toroidal, closed field line trap[3]. First results from this device were published earlier this year and are discussed further in this paper[4].

PHYSICS ISSUES

The goals of the program of plasma physics research at Lawrence University are to study equilibrium, stability, long time-scale dynamics, and limitations on transport for electron plasmas confined with purely toroidal magnetic fields. In addition, we aim to provide research opportunities for capable undergraduate physics students.

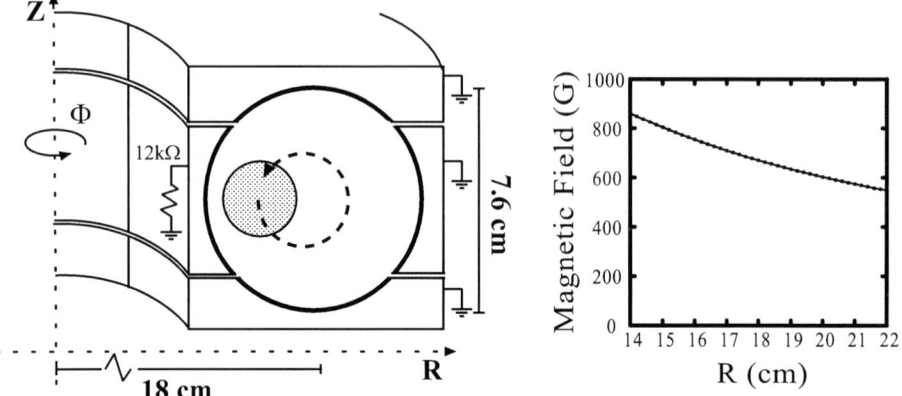

FIGURE 1. A depiction of the m=1 diocotron motion observed in the LNT II (left panel). The plasma (shaded circle) center executes a circular path (dashed line) about a point displaced toward the major axis and its motion induces corresponding oscillation of image charge on each wall segment. The right hand panel indicates the variation in the magnitude of the toroidal magnetic field across the minor diameter of the confinement region.

The existence of a stable equilibrium was predicted theoretically[5]. The stability argument rests on the adiabatic invariance of the single-particle magnetic moments and (toroidal) angular momenta as the particles $\mathbf{E} \times \mathbf{B}$ drift poloidally through the curved, non-uniform toroidal magnetic field. The variation in magnetic field strength across the poloidal cross-section of LNT II is shown in the right hand panel of Fig. 1. Although the equilibrium is a state of maximum potential energy, the kinetic energy is constrained by the invariants and the plasma cannot move away from the maximum energy state in the absence of some extraneous effect that permits energy transfer from the electrons (*e.g.* wall resistance or ion contamination). The poloidal $\mathbf{E} \times \mathbf{B}$ drift acts as an effective rotational transform that closes the single-particle drift orbits despite the presence of the vertical curvature and ∇B drifts. Since the electric field is due to the plasma space charge and the existence of the equilibrium relies on poloidal rotation, charge must be injected into the trap rapidly compared to the vertical drift timescale.

For the high aspect ratio toroidal electron plasmas studied at Lawrence University, the dynamics (*e.g.* the diocotron modes) are expected to be very similar to the dynamics of cylindrical electron plasmas. Subtle toroidal effects will be examined by a combination of experiment and numerical simulation.

Unlike cylindrical non-neutral plasmas for which true thermodynamic equilibrium exists and has been approached experimentally[6, 7], theory predicts a fundamental limit to the confinement of toroidal non-neutral plasmas[8]. Poloidal rotation of the plasma through the non-uniform toroidal magnetic field leads to magnetic pumping. Differential oscillations in the parallel and perpendicular temperatures (due to invariance of single-particle magnetic moment and toroidal angular momentum) combined with electron-electron collisions leads to irreversible conversion of electrostatic potential energy to thermal energy. Energy conservation requires expansion (*i.e.* transport) of the plasma.

One of the aims of our research program is to identify this theoretically predicted confinement limit. In order to do so, all other transport mechanisms must be reduced significantly compared to previous experiments. A scaling analysis of the magnetic pumping transport timescale, τ_{mp}, with experimental parameters leads to the following predicted scaling relation,

$$\tau_{mp} \approx 0.02 R_0 (\text{cm})^2 \sqrt{T(\text{eV})}, \qquad (1)$$

Where R_0 is the major radius and T is the electron temperature. Note that the predicted scaling relation is independent of magnetic field, plasma density or minor radius. The major radius of the LNT II plasma is about 17.4 cm and the temperature is estimated to be a few eV. The scaling relation therefore predicts a magnetic pumping transport timescale of order ten seconds.

EXPERIMENTAL DETAILS

The Lawrence Non-neutral Torus II was designed to provide improved vacuum conditions, enhanced magnetic field, and a higher degree of symmetry compared to the previous experiment. A metal-sealed, bakeable, low magnetic permeability stainless steel vacuum chamber is used to reduce or eliminate the effects of neutral gas. Experimental results presented here were obtained with base vacuum pressures as low as 8×10^{-9} Torr. A sixty turn, water-cooled, copper toroidal field coil, energized with 1 kA of D.C. current, generates a magnetic field strength approaching 700 gauss at the major radius ($R_0 \approx 17.4$ cm) of the plasma. The coil was designed to produce toroidal field ripple smaller than one part in 10^5, by symmetric arrangement of the conductors passing through the toroidal hole and by placing the twelve return leg bundles far from the plasma. A toroidally symmetric electrostatic boundary near the plasma is provided by a gold-plated aluminum shell. The shell is composed of trapezoidal modules that have been machined with a circular poloidal cross-section toroidal bore and cut into four (or eight) segments (see Fig.2). One of the modules (injection module, I in Fig. 2) has a slot in the outer wall into which a 1" diameter pancake spiral tungsten filament is inserted. The modules on either side of the injection module (G1 and G2) are employed as trapping electrodes. The remaining modules (C1-C5) form the 270° toroidal trapping region. Four of the five trapping region modules are divided into four electrically isolated segments (inner, outer, top and bottom). The isolated segments function as either wall-probe diagnostics for monitoring the image charge induced on the wall, or as antennae for launching plasma waves. One of the trapping region modules (C3) is divided into eight segments and can be used to observe higher order modes or to study the effect of a 'rotating wall' [9, 10]. The filament is directly heated with 11.5 A and has a radial potential profile that is designed to match the target plasma space charge potential profile for a plasma with an average density of 10^7 cm^{-3}. Filament emission (≈ 1 mA) is sufficient to fill the trap in about a microsecond, which is much faster than the timescale for the curvature and ∇B drifts to carry electrons out of the trap vertically ($\approx 500 \mu s$), ensuring rapid establishment of the space charge necessary to reach an equilibrium. The experiment is operated in a fill-trap sequence. Electrons are emitted from the tungsten filament directly onto the magnetic field lines. The G1 electrode is maintained at -50 V DC. A

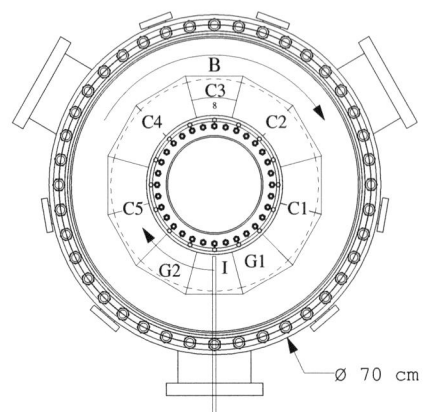

FIGURE 2. Top view schematic of the LNT II vacuum chamber and conducting shell. The module labeled I, is the site where electrons are injected from the tungsten filament. Modules labeled G1 and G2 serve as the gate electrodes. The modules labeled C1 - C5 make up the confinement region.

gated -50 V is applied to the G2 electrode to alternately fill (when G2 is grounded) and trap (when the potential on G2 goes negative) the plasma. Additional details about LNT II can be found in Ref. [3].

RESULTS

Diagnostic data is acquired in LNT II by monitoring the flow of image charge to and from isolated sectors of the conducting boundary. When a sector is used as a wall probe, the voltage across a 12 kΩ resistor (in parallel with the $\sim$1 nF capacitance to ground) is sampled. All wall sectors, except those being used as gate electrodes (G1 and G2 in Fig. 2) and those used as antennae for launching diocotron waves (see below), are maintained at ground potential by a low resistance connection.

A toroidal version of the $m = 1$ diocotron mode is observed in LNT II. The data shown in Fig. 3 are the simultaneously acquired signals from the four sectors on the C2 module (inner, top, outer, and bottom sectors in sequence from the top to the bottom panels). The dominant feature of this signal is a sinusoidal wave at approximately 50 kHz. The phase relation between the signals on the four electrodes is consistent with a poloidal mode number $m = 1$ wave. The phase difference is zero between signals sampled on toroidally separated electrodes at the same poloidal angle (not shown). The wave is therefore identified as the $m = 1$ diocotron mode. Figure 1 (left) depicts the plasma motion in the poloidal plane.

In the best vacuum conditions (and when a single wall sector is sampled to reduce dissipation of image currents) the $m = 1$ mode is not self-excited, but must be launched by application of five cycles (4 V_{pp}) at a frequency that is close to the mode frequency. Figure 4 shows the signal on the inner segment of the C2 electrode module when the mode was launched (using the inner segment of the C5 electrode module as the launch

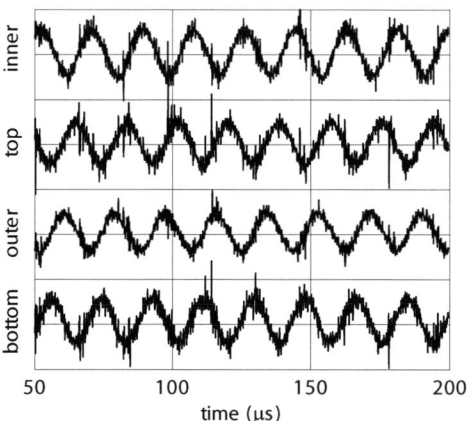

FIGURE 3. Wall probe signals from the four segments of the C2 electrode: inner, top, outer, bottom, sampled early in the trapping phase[4].

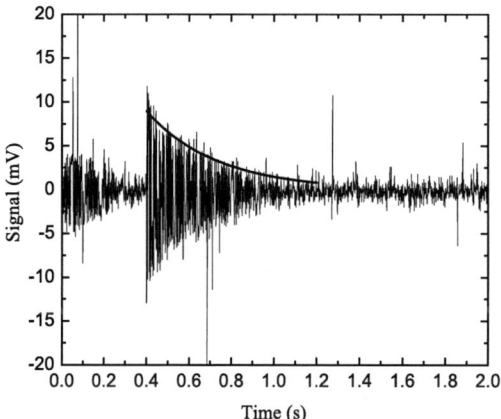

FIGURE 4. Wall probe signal from the inner segment of the C2 electrode module for the cases where the $m = 1$ mode was launched at 0.4 s after initiation of the trapping phase. The solid curve is an exponential decay with a 330 ms time constant.

antenna) about 0.4 second after the start of the trapping phase of the experiment. The $m = 1$ mode persists long after the excitation cycles are terminated and is damped on a $\sim$ 330 ms timescale. By repeating the process used to obtain the data in Fig. 4 with different delay times, the $m = 1$ mode frequency versus time can be measured. Since there is a proportional relationship between the $m = 1$ frequency and the trapped charge (for the mode in a cylindrical trap) [11], the plasma confinement time can be determined from the decay of the mode frequency. The theoretical expression for the small amplitude $m = 1$ mode in a cylinder is used to calculate the trapped charge per unit length and is indicated by the filled symbols in Fig. 5. Fitting an exponential decay to the data results

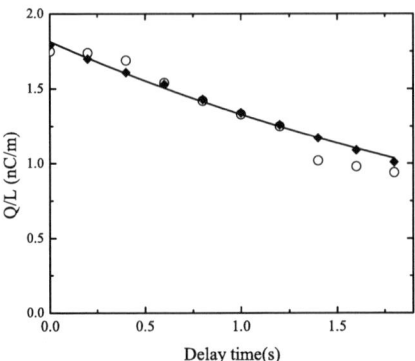

FIGURE 5. The trapped charge per unit length (along the magnetic field) inferred from the measured mode frequency using the theoretical relation (filled) and using a simulation that captures toroidal effects (open). The line is an exponential fit to the filled data points with decay constant of 3.2 ± 0.1 s.

in a charge confinement time of 3.2 ± 0.1 s.

SIMULATIONS

A complementary part of this project has been to employ numerical methods to simulate and study the toroidal effects on the $m = 1$ diocotron mode for an electron plasma. The conclusions discussed above rely on the toroidal dynamics being qualitatively similar to the cylindrical dynamics. Numerical simulations that include toroidal effects validate this assumption for the large aspect ratio LNT II device.

Toroidal effects on the $m = 1$ diocotron mode are simulated for an axisymmetric, uniform density plasma. Poisson's equation is solved in the poloidal plane (cylindrical coordinates R and z with the azimuthal angle, ϕ, being a direction of symmetry, see Fig. 1) using a relaxation algorithm on a uniform grid with a grounded boundary corresponding to the experimental boundary. The input parameters are the total charge, plasma radius, and the initial position of the plasma center. The charge on the sections of the conducting boundary is determined by integrating the normal component of the electric field (obtained from the Poisson solution). By evaluating the $\mathbf{E} \times \mathbf{B}$ drift velocity at the plasma center, the plasma position is updated and Poisson's equation solved again. The charge on each wall sector is stored and a simulated image current signal generated by taking a numerical derivative and using the known experimental impedance (dominantly capacitive).

The simulation was run for a range of input parameters and two primary signal characteristics were determined from the power spectrum for comparison with experimental signal characteristics. The power weighted average frequency of the $m = 1$ mode and the (power weighted average) amplitude of the second harmonic normalized to the amplitude of the fundamental were found to correlate uniquely with the values of the input

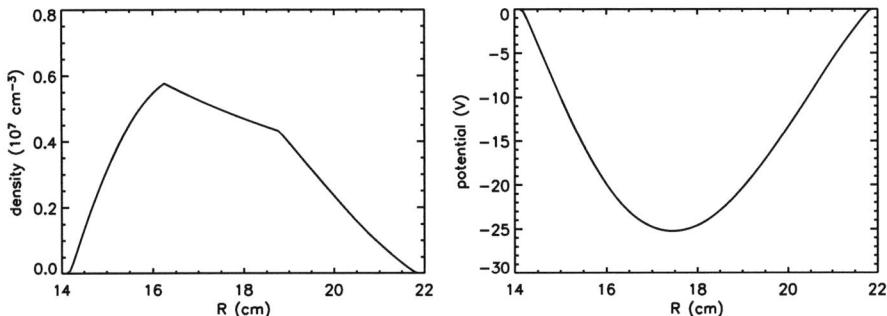

FIGURE 6. The electron density (left) and electric potential (right) along the midplane for the equilibrium solution to the Daugherty-Levy equation with Q/L=1.8 nC/m

parameters. By calculating the $m = 1$ frequency and the normalized second harmonic amplitude from the experimentally measured signal, the total charge in the plasma as well as the trajectory of the plasma center could be determined by interpolating on the grid correlating simulation signal characteristics to input parameters. The trapped charge per unit length versus time thus determined, shown in Fig. 5 (open symbols), is nearly identical to the values obtained using the cylindrical theory (filled symbols). We therefore observe no significant toroidal modification to the $m = 1$ mode frequency. The trajectory of the center of the plasma is centered on a point that is displaced about 5 mm from the geometric center of the conducting boundary. This inward shift of the equilibrium position is a toroidal effect that is analogous (although in the opposite direction) to the Shafranov shift for toroidal MHD plasmas. The simulation also provides a measure of the amplitude of the $m = 1$ motion of the center of charge. In the cases shown, the amplitude is about a millimeter and presumably depends on how precisely the launch frequency matched the natural frequency.

DISCUSSION AND CONCLUSIONS

The magnetic pumping transport mechanism is expected to limit confinement in the LNT II to about ten seconds (Eq.1). The measured confinement time approaches this theoretical limit and suggests that we may be able to identify this transport mechanism in future experiments.

The confinement time is also to be compared to the $\mathbf{E} \times \mathbf{B}$ rotation period in order to demonstrate the conclusion that nearly steady-state conditions are achieved. Efforts to measure the $m = 2$ mode are underway and will provide an experimental measure of the electron density. Absent those measurements we resort to equilibrium modeling in order to estimate the density. The Daugherty-Levy equation governs the equilibrium for cold toroidal non-neutral plasmas[12]. We solve this nonlinear partial differential equation numerically, subject to the experimental constraints that the charge per unit length be 1.8 nC/m (Fig. 5) and that the central space charge potential be more positive than the central filament potential ($\approx$ -27 V). A physically reasonable solution is realized with a (nominally) trapezoidal density profile that is $\approx 0.5 \times 10^7$ cm^{-3} in the plasma center and falls smoothly to zero at the conducting boundary outside a 1.25 cm radius region that corresponds to the size of the filament (see Fig 6). The $\mathbf{E} \times \mathbf{B}$ rotation frequency associate with the inferred density is $\approx$ 100 kHz. The three second confinement time therefore represents 3×10^5 rotation periods. Since the rotation period is the longest dynamical timescale associated with the equilibrium, we conclude that nearly steady state conditions are achieved in LNT II.

In conclusions, pure electron plasmas are produced in a 270° arc of a toroidal trap. In best vacuum conditions ($<10^{-8}$ Torr), the $m = 1$ mode is stable or damped. The charge loss time (i.e. confinement time) is measured to be $\sim$3 seconds. This represents an order of magnitude improvement over previous experiments and, being more than 10^5 $\mathbf{E} \times \mathbf{B}$ rotations, indicates the establishment of nearly steady-state conditions. The charge loss time is of the same order of magnitude as the expected magnetic pumping transport time.

Future efforts will be dedicated toward producing trapped plasmas in a full torus, identifying the magnetic pumping transport mechanism, and characterizing plasma dynamics in a full torus. The long confinement times demonstrated here should permit withdrawal of the filament and relaxation of the trapping gate potentials to reach this goal.

ACKNOWLEDGMENTS

The authors thank D. P. Ryan, J. O. Hector and S. K. Curry for their contributions to the LNT II project. This work is supported by the National Science Foundation Grant No. PHY0317412.

REFERENCES

1. M. R. Stoneking, P. W. Fontana, R. L. Sampson, and D. J. Theucks, *Physics of Plasmas* **9**, 766–771 (2002).
2. M. R. Stoneking, M. A. Growdon, M. L. Milne, and R. T. Peterson, *Physical Review Letters* **92**, 095003 (2004).
3. J. P. Marler, and M. R. Stoneking, "A kilogauss-scale, high-vacuum toroidal experiment," in *Non-neutral Plasma Physics VI*, edited by M. D. et al., AIP Conference Proceedings 862, American Institute of Physics, New York, 2006, pp. 71–77.
4. J. P. Marler, and M. R. Stoneking, *Physical Review Letters* **100**, 0155001 (2008).
5. T. M. O'Neil, and R. A. Smith, *Physics of Plasmas* **1**, 2430–2440 (1994).
6. C. F. Driscoll, J. H. Malmberg, and K. S. Fine, *Physical Review Letters* **60**, 1290 (1988).
7. T. M. O'Neil, and C. F. Driscoll, *Physics of Fluids* **22**, 266 (1979).
8. S. M. Crooks, and T. M. O'Neil, *Physics of Plasmas* **3**, 2533–2537 (1996).
9. F. Anderegg, E. M. Hollmann, and C. F. Driscoll, *Physical Review Letters* **81**, 4875–78 (1998).
10. X.-P. Huang, F. Anderegg, E. M. Hollmann, C. F. Driscoll, and T. M. O'Neil, *Physical Review Letters* **78**, 875–878 (1997).
11. R. C. Davidson, *Physics of Nonneutral Plasmas*, Addison-Wesley, Redwood City, CA, 1990.
12. J. D. Daugherty, and R. H. Levy, *Physics of Fluids* **10**, 155–161 (1967).

Recent Progress on Toroidal Nonneutral Plasmas Confined on Heliotron Magnetic Surfaces

H. Himura[a], K. Nakamura[a], D. Sugimoto[a], A. Sanpei[a], S. Masamune[a], M. Isobe[b], F. Sano[c]

[a]*Kyoto Institute of Technology, Department of Electronics, Kyoto 606-8585, Japan*
[b]*National Institute for Fusion Sciences, Toki 509-5292, Japan*
[c]*Kyoto University, Institute of Advanced Energy, Uji 611-0011, Japan*

Abstract. Firstly, non-constant space potential ϕ_s and electron density n_e on magnetic surfaces of helical nonneutral plasmas are observed in CHS experiments. The variation of ϕ_s has grown with increasing electron injection energy, implying that thermal effects are important when considering the force balance along magnetic field lines. These observations confirm the existence of plasma equilibrium having non-constant ϕ_s and n_e on magnetic surfaces of helical nonneutral plasmas. Secondly, outward orbits that extend to inward part of closed helical vacuum magnetic region are found in three dimensional calculations which take into account two experimental findings. The pitch angle of electron injected into the stochastic magnetic region is scattered considerably due to the presence of shifted self space potential ϕ_s. Eventually, the injected electron turns to be a helically trapped particle, and start an inward movement along one of the $|B_{min}|$ contours. Once penetrating deeply, the electron can never escape from the last closed flux surface because the negative ϕ_s acts as a potential barrier. Thirdly, first data obtained from toroidal nonneutral plasmas confined on the quasi-poloidally symmetric magnetic surfaces having a helical-axis is described. It shows that externally produced plasmas are successfully observed, and moreover, the non-constant ϕ_s and n_e of helical nonneutral plasmas are ensured in the Heliotron J experiments, as well.

Keywords: toroidal nonneutral plasma, helical magnetic surface, helically trapped particle
PACS: 52.65.Cc, 52.55.Hc, 52.27.Jt, 52.27.Aj

I. EQUILIBRIUM WITH POTENTIAL AND DENSITY GRADIENTS ALONG MAGNETIC SURFACES

In the equilibrium state of pure electron plasmas confined on magnetic surfaces, the force balance equation of the electron fluid is written as

$$m_e n_e \vec{v}_e \cdot \nabla \vec{v}_e = -e n_e \vec{v}_e \times \vec{B} + e n_e \nabla \phi_s - \nabla p_e, \qquad (1)$$

where m_e, v_e, and p_e stand for the electron mass, velocity field, and the pressure of the electron fluid, respectively. When n_e is far below the Brillouin density limit $n_B = e_0 B^2 / 2 m_e$, the convective term on the left hand side can be ignored. In addition, when the plasma is assumed to be cold ($T_e = 0$), the ∇p_e term also vanishes. In that case, Eq. (1) takes on the simple form of $\vec{v}_e \times \vec{B} = \nabla \phi_s$, which leads to $\vec{B} \cdot \nabla \phi_s = 0$. This means

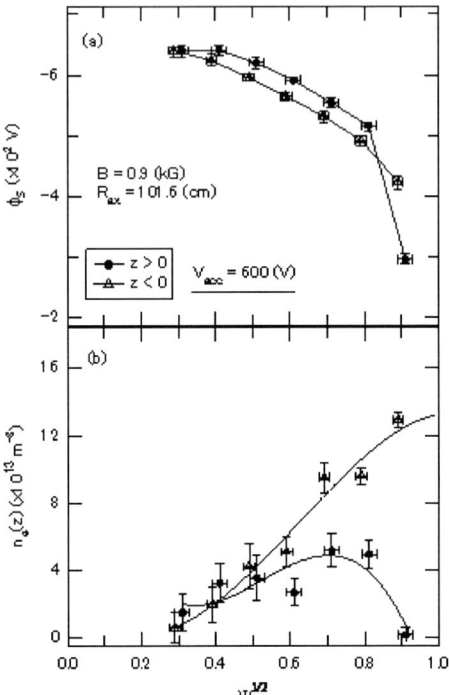

FIGURE 1. Profiles of (a) space potential $\phi_s(z)$ and (b) the inferred electron density $n_e(z)$ measured at the 6-U cross section for the case of $V_{acc} = -600$ V. The position of the e-gun is at 1 cm outside the LCFS on a different poloidal cross section (see Ref. [2].)

that ϕ_s must be constant along magnetic field lines of force, that is, on the magnetic surfaces. However, when T_e is finite and p_e is relatively low, that is not the case. In this case, we must consider the parallel component of the force balance that is written as

$$en_e \nabla_\| \phi_s (=-en_e E_\|) = \nabla_\| p_e, \qquad (2)$$

where $\nabla_\|$ represents the derivative along the magnetic field line of force. Assuming that T_e is constant on each magnetic surface, that is, T_e is a function of magnetic flux Ψ, Eq. (2) can be rewritten as

$$en_e \nabla_\| \phi_s (=-en_e E_\|) = T_e(\Psi) \nabla_\| n_e. \qquad (3)$$

Equation (3) can be integrated and has the solution $n_e \sim \exp(e\phi_s / T_e(\Psi))$ if n_e need not be constant along the magnetic field line of force. On the other hand, ϕ_s and n_e are also related to each other as $\nabla^2 \phi_s = en_e / \varepsilon_0$, where ε_0 is the permittivity of free space. From the above consideration, the Poisson-Boltzmann equation has been derived for the equilibrium of one-species plasmas confined on magnetic surfaces [1]. And, Eq. (3) predicts that n_e and consequently ϕ_s need not be magnetic surface quantities when the electrostatic force is not small compared to the pressure force. One of key parameters of the theory is a/λ_D, where a and λ_D are a typical length of plasmas (~ average radius) and the Debye length, respectively. In fact, for $a/\lambda_D \sim 1$-10, it is expected that ϕ_s and n_e vary significantly on magnetic surfaces.

The above prediction has been verified by the experiments on the Compact Helical System (CHS) device, for the first time [2]. Fig. 1 (a) shows a typical profile of space potential $\phi_s(z)$ measured by the high-impedance emissive method. Here, $\Psi^{1/2} = 0$ and 1 correspond to the magnetic axis R_{ax} and the last closed flux surface (LCFS), respectively. Differences between two values of ϕ_s at $z > 0$ (henceforth ϕ_{s_u}) and ϕ_s at $z < 0$ (henceforth ϕ_{s_d}) on each magnetic surface (at same value of $\Psi^{1/2}$) are observed. This means that ϕ_s is not constant on magnetic surfaces, contrary to the equilibria of fusion plasmas. As is shown from the data, the difference in ϕ_s becomes larger in the outer region of magnetic surfaces. Actually, the difference reaches about 120 V at $\Psi^{1/2} \sim 0.9$, while at $\Psi^{1/2} \sim 0.3$ the difference almost disappears. Also, one notes that there are two "*crossover*" points of the measured $\phi_s(z)$, which are at $\Psi^{1/2} \sim 0.3$ and 0.8. Inside $\Psi^{1/2} \sim 0.3$, ϕ_{s_u} seems to be somewhat smaller than ϕ_{s_d} or almost equal to it, while $\phi_{s\_u} > \phi_{s\_d}$ between the two crossover points. Outside $\Psi^{1/2} \sim 0.8$, the magnitude relation is flipped again, that is, $\phi_{s\_u} < \phi_{s\_d}$. The corresponding profile of n_e is clearly observed, as shown in Fig. 1 (b). Readers can refer to Ref. 2 in order to know the details of this topic.

II. PENETRATION OF HELICALLY TRAPPED ELECTRONS ACROSS SEPARATRIX

In experiments on CHS and also Heliotron J [3], an electron-gun (hereafter, e-gun) has been installed in the stochastic magnetic region (SMR) surrounding the last closed flux surface (LCFS) and just ejected thermal electrons in the SMR. Then, within the order of 10 μs after the injection, those have penetrated deeply in the helical magnetic surfaces (HMS), spread rapidly in the whole of the closed surfaces, and finally formed a helical nonneutral plasma there. Since the observed penetration [4] has happened in much shorter than any classical binary collision time, some cross-field transport associated with free-streaming of electrons along the stochastically wandering field lines had been considered as a possible mechanism. However, no electron orbit that extended to the innermost region of HMS was found [5] in three dimensional numerical calculations of a single electron motion in CHS vacuum magnetic fields.

There exist two experimental findings that have remained to be excluded in the past calculations. One is the effect of space potential ϕ_s in the SMR, which has been discovered in the not only CHS nonneutral- [4] but also neutral experiments [6]. Actually, ϕ_s attributes to the bunch of electrons confined in the SMR so that ϕ_s can be sorted as self electric potential. The other is the fact that ϕ_s varies on the HMS, in other words, values of ϕ_s are nonconstant along magnetic field lines of helical nonneutral plasmas, as explained in Sec. I. Obviously, these proofs would affect on the electron orbit, which calls for a renewed calculation including them.

Figure 2 shows a typical set of plasma data of a single electron, which are obtained under the model where the two findings are taken into account. Considerable propaga-

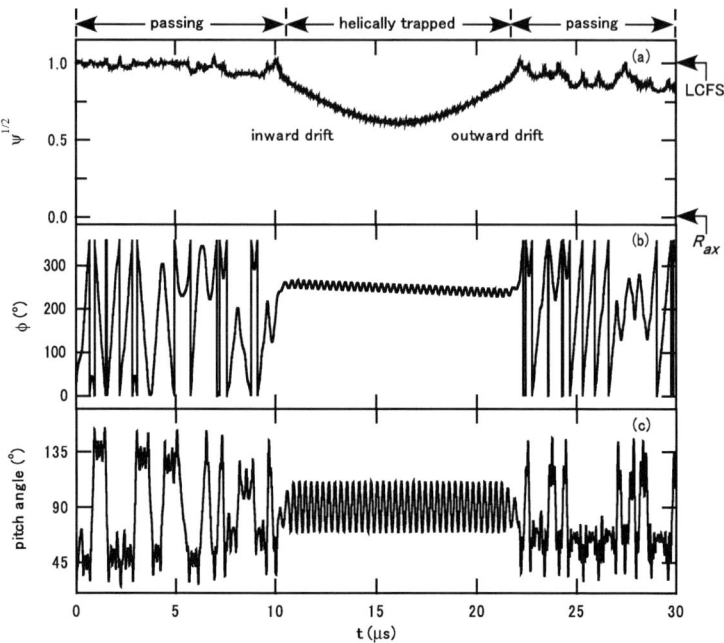

FIGURE 2. Typical time evolutions of (a) normalized position, (b) toroidal angle, and (c) pitch angle of the electron injected into the SMR. The detail of the calculation is described in Ref. [9].

tion across the HMS is observed. At $t = 0$ μs, the electron is injected into the SMR. Since the position is represented using $\Psi^{1/2}$, the period of $\Psi^{1/2} > 1$ means that the injected electron is circulating in the SMR, in other words, outside the LCFS. On the other hand, when $\Psi^{1/2} < 1$, the electron is inside the LCFS, which thus indicates that the inward penetration of the electron has happened. In fact, such penetration can be clearly recognized for the period of $t > 10.4$ μs. Looking at all time evolutions carefully, one recognizes what steps the electron follows in order to propagate across the HMS. Firstly, as seen from Fig. 2 (b), just after the injection (at $t = 0$ μs), the electron immediately rounds the torus within ~ 0.8 μs. This can be recognized from the fact that the toroidal angle ϕ starts to increase almost linearly from 50° at $t = 0$ μs and returns to there at $t \sim 0.8$ μs. This also means that the electron rotates counter-clockwise. Then, the value of ϕ decreases suddenly, indicating that the electron has initiated to co-rotate in toroidal direction. These motions have been referred as 'passing particle [7]'. As recognized from Fig. 2 (c), this event is triggered by the considerable change in the pitch angle of the electron, which is apparently due to the presence of ϕ_s. Similar motion like the passing particle lasts until $t \sim 10.4$ μs, although the electron sometimes experiences magnetic mirror reflection. During the period, the electron cannot penetrate deeply in the HMS. Only slight drifting due to both $\mathbf{E_s \times B}$ and ∇B drifts is observed, as seen in Fig. 3 (a). However, a significant change in the motion occurs at $t \sim 10.4$ μs. After the time, the electron starts a movement to propagate in the HMS and quickly arrive at $\Psi^{1/2} \sim 0.6$ within only ~ 6 μs. It should be

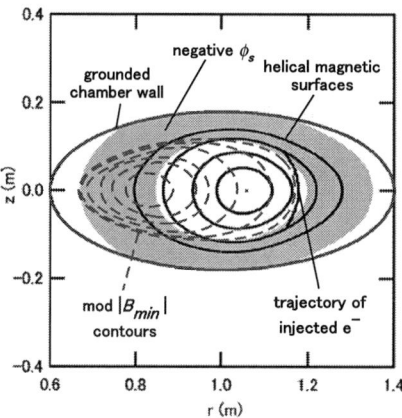

FIGURE 3. The trajectory of the helically trapped electron projected onto the $\phi = 247.5°$ poloidal plane where both the $|B_{min}|$ contours and the HMS are described. The detail can be found in Ref. [9].

noted again that no ϕ_s exist at $\Psi^{1/2} < 0.8$. Nevertheless, the electron propagates in the region across the HMS. Then for $t > 16$ μs, the electron drifts outwardly toward the outermost region of HMS. When the inward penetration occurs, considerable changes also appear in time histories of both ϕ and the pitch angle. As clearly seen from the data in Fig. 2 (b), the electron is completely trapped in the vicinity of $\phi \sim 247.5°$. Meanwhile, the pitch angle changes periodically between 70° and 110°, as seen in Fig. 2 (c). These results indicate that the electron turns to be a 'helically trapped particle [8]' when it is propagating across the HMS.

It is well-known that once a charged particle is trapped in the bottom of the ripple of B-field strength, the trapped particle can easily drift across the HMS. Since the bottom of the ripple is the region of weaker B-field, drift surfaces for the orbit of the helically trapped particle are thus almost coincided with the contour plots of the weaker field (called the $|B_{min}|$ contours [7]). Figure 3 shows projections of the contours and the electron trajectory onto the $\phi = 247.5°$ poloidal plane. As recognized, the contours never coincide with the HMS. Comparing the trajectory of the electron with the $|B_{min}|$ contours, it can be actually found that the trajectory is approximately along one of contour curves (see also Fig. 3 (a)) for this case. Therefore, we have concluded that the observed inward penetration is caused by the drift motion of helically trapped particle. Another fact which supports this conclusion is that the injected electron has always drifted from the upper ($z > 0$) to the downside ($z < 0$) region in the LCFS, when it starts the drift motion. This direction is actually consistent with that of the ∇B drift expected for electrons. And, if this inward penetration happens continuously for some of the following electrons launched from the e-gun, the closed magnetic region would be filled up with them from the outermost part and finally a helical nonneutral plasma would be formed. The detail of this work is now under reviewed [9]. Finally, it should be briefly mentioned why helically trapped electrons can be accumulated in the HMS. In this presented case, finite negative ϕ_s due to the space charge of launched electrons has been extended to the SMR and its vicinity. The helically trapped electron can actually propagate from the outward negative ϕ_s region to the inward vacuum region, whereas the electron drifting outwardly is reflected at the negative ϕ_s region (see also $t > 22$ μs in Fig. 3). In other words, the negative ϕ_s region acts as a *potential barrier*. Such a reflection is never the case for conventional helical plasmas where no negative ϕ_s exists in the SMR and its vicinity, which causes helically trapped electrons escape out from the LCFS.

III. NEW EXPERIMENT ON MAGNETIC SURFACES WITH A HELICAL MAGNETIC AXIS: HELIOTRON J

While CHS is one of helical devices with a plane-axis, there exist other devices having a helical-axis. We have performed a preliminary experiment on a helical-axis device called Heliotron J [3]. Contrary to the CHS device, it forms quasi-poloidally symmetrical magnetic surfaces and the *B*-field strength is almost constant at the central cross section of the straight part of the device. Thus, this property may result in producing quasi-isodynamic plasmas with a longer confinement time than those of CHS nonneutral experiments.

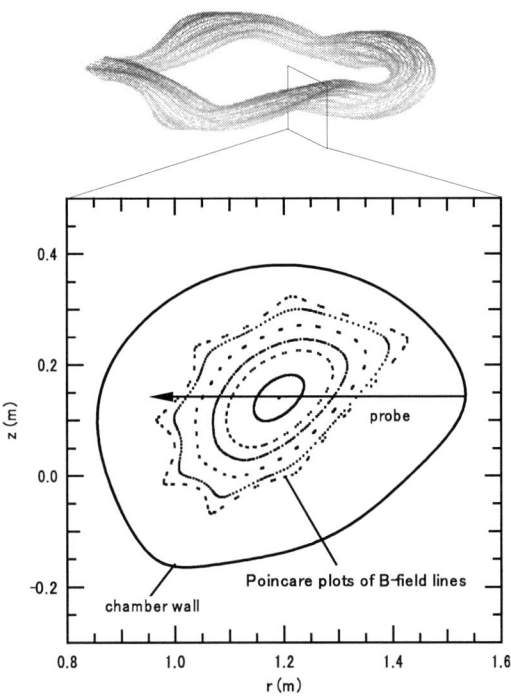

FIGURE 4. A schematic drawing of three-dimensional structure of magnetic surfaces of Heliotron J having a helical-axis, and the magnetic surfaces projected on the a poloidal cross section.

TABLE 1. Principal parameters of helical nonneutral experiments on Heliotron J

	Plasma parameter	Value
Heliotron-J	Major radius	1.2 m
	Averaged minor radius	0.38 m
	Pole number	1
	Field period number	4
	Field strength on axis	~0.3 kG
Electron gun (e-gun)	Emission area	~1 cm^2
	Acceleration	~600 V
	Beam current	<0.1 A
	Beam density	$<1 \times 10^{14}$ m^{-3}
Helical electron plasma	Averaged volume density	$\sim 2 \times 10^{10}$ m^{-3}
	Temperature	<180 eV
	Electron Larmor radius	~3.8 mm
	Maximum space potential	~500 V

A. Experimental setup

A schematic drawing of the three-dimensional structure of the Heliotron J vacuum magnetic surfaces with a poloidal cross sectional view is described in Fig. 4 and principal parameters of experiments are listed in Table I. The B-field is static and the maximum value of B at R_{ax} is about 0.3 kG. Typical vacuum pressure of Heliotron J is about 2×10^{-9} Torr. In order to produce electron plasmas, we use the same LaB$_6$ emitter as we employed in CHS experiments. The electron gun with the emitter is inserted horizontally along the r axis. Electrons can be launched out into the vacuum B field with V_{acc} up to ~ - 600 V. The emission beam current I_b is variable, but we have limited I_b to ~ 0.1 A. A Langmuir emissive probe is used to measure ϕ_s and the probe current I_p in helical electron plasmas. To prevent short circuiting of ϕ_s across magnetic surfaces, both the probe shaft is covered with a quartz tube; the outer diameter of the tube is 6 mm. The probe is inserted horizontally along the r axis, as seen in Fig. 4.

B. Preliminary results

In Fig. 5, profiles of ϕ_s and n_e of helical nonneutral plasmas confined on magnetic surfaces of Heliotron J are described. As shown from the plotted data in Fig. 5 (a), substantial differences between two values of ϕ_s measured in the outboard- and inboard side on each magnetic surface (at the same value of $\Psi^{1/2}$) are observed. This means that ϕ_s is not constant on magnetic surfaces. As also recognized, the difference in ϕ_s becomes larger in the outer region of magnetic surfaces. Actually, the difference reaches about 300 V at the LCFS, while at $\Psi^{1/2} \sim 0.4$ the difference almost disappears.

Similar non-constant appears also in the measured n_e, as recognized in Fig. 5 (b). The plotted data indicate that values of n_e in the inboard side are larger than those in the outboard one. This tendency is similar to the observed ϕ_s (see also Fig. 5 (a)), but reverse of that in the CHS experiments. Also, values of n_e seem to be too small compared to the CHS experiments, despite the same B-field strength with the order 0.1

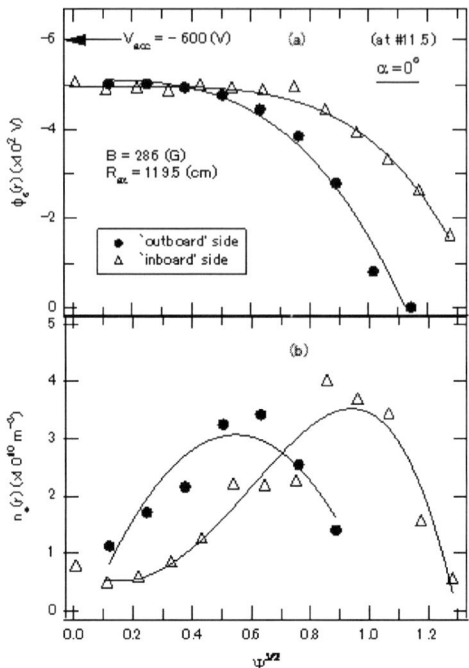

FIGURE 5. Preliminary data (a) space potential $\phi_s(r)$ and (b) the inferred electron density $n_e(r)$ measured on the Heliotron J device having a helical magnetic axis.

kG. Actually, this small value of n_e has resulted in long Debye length, suggesting no fluid approximation is held for the Heliotron J nonneutral plasmas. More data are thus required to investigate the plasmas experimentally. The second series of experiments is scheduled in this fall.

ACKNOWLEDGMENTS

This work is performed under the auspices of the NIFS CHS Research Collaboration (NIFS07KZPH002) and the NIFS Heliotron J Research Collaboration (NIFS08KUHL018).

1. T. S. Pedersen and A. H. Boozer, Phys. Rev. Lett. **88**, 205002 (2002).
2. H. Himura, H. Wakabayashi, Y. Yamamoto et al., Phys. Plasmas **14**, 022507 (2007).
3. T. Obiki, F. Sano, M. Wakatani et al., Plasma Phys. Control. Fusion **42**, 1151-1164 (2000).
4. H. Himura, H. Wakabayashi, M. Fukao et al., Phys. Plasmas 11, **492** (2004).
5. H. Wakabayashi, "Experimental studies on producing toroidal nonneutral plasmas on helical magnetic surfaces on CHS", Ph.D. Thesis, The Graduate University for Advanced Studies, 2006.
6. E. D. Volkov, V. L. Berezhnyi, V. N. Bondarenko et al., Czechoslovak Journal of Physics **53**, 887-894 (2003).
7. for example, M. Isobe, M. Sasao, S. Okamura et al., Nucl. Fusion **41**, 1273-1281 (2001).
8. for example, T. Watanabe, Y. Matsumoto, M. Hishiki et al., Nucl. Fusion **46**, 291-305 (2006).
9. K. Nakamura, H. Himura, M. Isobe et al., submitted to Phys. Plasmas (2008).

Confinement of pure electron plasmas in the CNT stellarator

T. Sunn Pedersen*, J. W. Berkery*, A. H. Boozer*, Q. R. Marksteiner*,[†], P. W. Brenner*, M. Hahn*, B. Durand de Gevigney* and X. Sarasola Martin*

*Columbia University, New York, NY 10027
[†]Los Alamos National Laboratory, Los Alamos, NM 87545

Abstract. The Columbia Non-neutral Torus is a stellarator devoted to non-neutral and electron-positron plasma research. Confinement and transport processes have been studied in some detail now, and an understanding of these processes has emerged. Transport is driven in two ways: The presence of internal rods, and the presence of neutrals. Both transport processes are clearly distinguished experimentally, and a model of the rod driven transport has been developed, yielding very good agreement with experimental data. The neutral driven transport is faster than originally expected and indicates the presence of unconfined orbits in CNT. Numerical modeling of the electron orbits in CNT confirms the existence of loss orbits and shows that a flux surface conforming electrostatic boundary will greatly improve confinement. Such a boundary has now been installed in CNT, with initial results showing an order of magnitude improvement in confinement.

Keywords: Non-neutral plasmas, pure electron plasmas, plasma confinement, stellarators, magnetic confinement, neoclassical transport, electron-positron plasmas
PACS: 52.27.Jt,52.27.Ep,52.27.Aj

INTRODUCTION

The Columbia Non-neutral Torus is a compact, simple stellarator [1, 2] devoted to the studies of non-neutral and electron-positron plasmas on magnetic surfaces. Pure electron plasmas confined in a stellarator are predicted to be fundamentally different from pure electron plasmas in Penning traps [3] and pure toroidal field traps [4] because the equilibrium on each magnetic surface is a minimum energy state [5], rather than a maximum energy state [6]. The confinement of a pure electron plasma is predicted to be much better than that of a quasineutral plasma [7]. If confinement indeed can be made very long, it may be possible to create electron-positron plasmas in CNT with existing positron sources [8]. If a nearly instantaneous source of $> 10^{11}$ cold ($T_p < 5$ eV) positrons becomes available [9], then the electron plasma confinement time would only need to be somewhat better than 20 msec, the retraction time of the electron emitter in CNT [10]. This less stringent confinement time has now been achieved in CNT, as will be discussed in this article.

CREATION AND DIAGNOSIS OF CNT PLASMAS

Injection of electrons

For all experiments discussed here, electrons were injected near the magnetic axis (ie. the toroidal center line of the confinement region) by thermionic emission from a heated, negatively biased, stationary filament. Parallel transport fills up the central region with electrons in a few μs, and then the net radial outward transport fills the magnetic surfaces. The experiments discussed here are in steady state between the emission at the magnetic axis and the radial losses at the last closed magnetic flux surface.

Diagnostic techniques

These steady state pure electron plasmas are diagnosed using internal particle probes, operated either as regular Langmuir probes, ie. non-emissive collection probes, or as emissive probes. Current-voltage characteristics of these probes are used to determine the equilibrium electron plasma density, temperature, and electrostatic potential profiles [11]. In particular, one should note that in these extremely low density plasmas, the most reliable way to measure the space potential is to use deviation potential method rather than using the floating potential of high impedance probes.

ROD DRIVEN TRANSPORT

Experimental observations

The first transport process to be unambiguously identified, characterized and modeled, was the transport driven by the internal rods in CNT. As mentioned, diagnostic probes and emitters are inserted into the plasma. These are mounted on rods made of an insulating ceramic material, alumina, and are therefore not direct steady state sinks or sources of electrons. However, they are still observed to drive radial transport. At low neutral pressures ($p_n < 10^{-8}$ Torr), the rods actually dominate the radial losses, as was verified experimentally [12]. Despite these losses, the confinement time in CNT was observed to be as large as 20 msec.

Numerical and theoretical modeling of rod-driven transport

It is the electrostatic perturbation created by the negatively charged self-shielding rod that causes these losses. A relatively simple model for the rod driven transport was derived. The model includes the radial ExB losses created by the electric field of the rod, with Debye screening included. There is good agreement between the model and the experimental data [13] although the model systematically underestimates the transport rates. This is likely due to the Debye screening model, which assumes that the electric field drops off exponentially with a characteristic scale length equal to the Debye length in CNT, typically 1.5 cm. However, Debye screening is rather a subtle process in a pure electron plasma; in the absence of ions, a strongly negatively charged object will repel all electrons and create a vacuum region around itself. This region will extend for several Debye lengths, and the electric field will not drop off exponentially in this region. Thus the region of perturbed electric field is larger in the experiment than what is assumed in the model. It should also be noted that the rod driven transport depends on the magnitude of the Debye length, see the Discussion and Conclusion Section.

NEUTRAL DRIVEN TRANSPORT

Neutrals also drive transport in CNT. The loss rate shows an offset linear relationship with the neutral pressure, with the offset (finite transport at a perfect vacuum level) attributed to the aforementioned rod driven transport. By varying the neutral pressure in experiments, one can separate the neutral and rod driven transport and study the two separately. The study of neutral driven transport is of particular interest in CNT, since electron-neutral collisions are relatively easy to model and compare to neoclassical predictions. Electron-neutral collisions are large pitch angle scattering collisions with almost no energy scattering. The lack of energy scattering is because of the large mass of the neutral relative to the electron, and the pitch angle scattering is a simple classical collision where the magnetic field can be neglected, since the electron Larmor radius is orders of magnitude larger than the neutral atom radius. In the following, we will briefly summarize the main results from studies of neutral driven transport. More details can be found in Ref. [13]

B-field scaling

When increasing the B-field strength, we observe a decrease in transport Γ that can be modeled as $\Gamma \approx \Gamma_0 + \alpha * B^{-1.5}$, with Γ_0 and α constants. That is, there are apparently two transport mechanisms driven by the neutral collisions, one that is independent of the B-field, and one that depends on the B-field strength as $B^{-1.5}$, Fig. 1.

Evidence of prompt orbit losses

Both of the mentioned scalings strongly suggest that there are unconfined orbits in CNT. The B-field independent scaling is indicative of a class of unconfined orbits that stay unconfined even as the magnetic field is increased. Such orbits are generically found in magnetic mirror machines and in unoptimized stellarators. Although CNT is also an unoptimized stellarator, the strong electric field was predicted to close the orbits of the magnetically trapped particles, which will otherwise leave on the ΓB drift time scale, which is on the order of 10^{-5} seconds, essentially instantaneous compared to the CNT confinement time $10^{-3} - 10^{-1}$ seconds. Raising the magnetic field will increase the ΓB drift time but it remains negligible compared to the confinement time in CNT, so

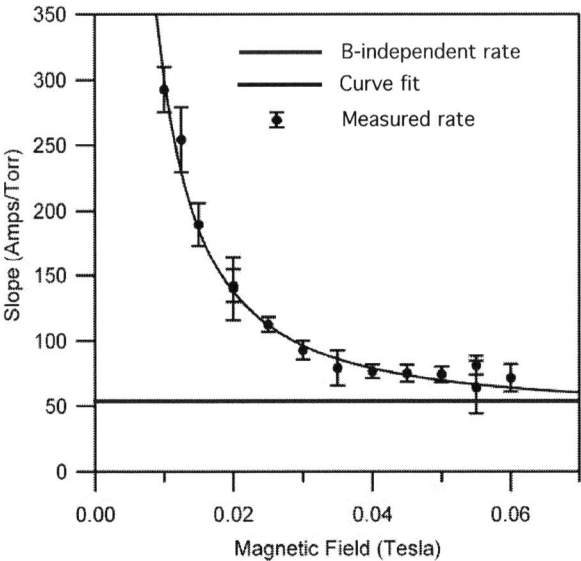

FIGURE 1. The neutral driven transport rate, expressed in A/Torr, compared with the curve fit of a $B^{-1.5}$ scaling and a B-field independent scaling. The B-field independent transport rate is noted on the figure with a blue line

it appears experimentally as if this transport mechanism is independent of B. Instead, the transport rate will scale with the collision frequency, since collisions can change the particle's orbit from a confined orbit to an unconfined orbit, after which it leaves the plasma.

The $B^{-1.5}$ scaling indicates that some orbits do transition from unconfined to confined as the magnetic field is raised. This scaling has also been observed in Penning traps [14] and the pure toroidal field trap [15], and is believed to be due to trapped particles whose orbits may change from open (loss orbits) to closed (confined orbits) if the magnetic field is sufficiently strong.

Using the aforementioned method of separating the rod driven transport, and the neutral driven transport, we also see that if we ignore the rod driven transport, the confinement time is essentially one collision time. This is further indication that there are unconfined orbits in CNT, and that a substantial part of orbits in CNT are unconfined.

NUMERICAL STUDY OF PARTICLE ORBITS IN CNT

The experimental observations of the neutral driven transport show that there are unconfined orbits in CNT. This is somewhat unexpected since the electric field is strong, causing poloidal rotation of magnetically trapped particles. A numerical study of the orbits in CNT has been performed, and is described in detail in another contribution at this conference [16]. The main results are briefly summarized here. It is found that indeed there are unconfined orbits in CNT, caused by the electrostatic boundary condition that was used until fall 2007. This boundary condition was set by the grounded casings of the two internal, interlocking coils, and to a lesser degree, by the much more distant grounded vacuum chamber. This imposed strong variations in electrostatic potential on the outer magnetic surfaces. These strong variations imply that the $E \times B$ drift has a component across the magnetic surfaces, allowing the particle to drift to an outer magnetic surface. The particle can then move along the magnetic field to another electrostatic potential surface, and can $E \times B$ drift further out. A large fraction of electron orbits on the outer magnetic surfaces is unconfined due to the substantial misalignment between the magnetic surfaces and the surfaces of constant electrostatic potential. If the electrostatic potential is assumed to be constant on a magnetic surface, the simulations show that indeed the electrostatic potential and its associated $E \times B$ rotation, is beneficial for confinement and essentially all particle orbits are confined.

IMPROVED CONFINEMENT WITH CONFORMING ELECTROSTATIC BOUNDARY INSTALLED

The numerical findings clearly show the importance of having an electrostatic boundary condition that minimizes electrostatic potential variations on the magnetic surfaces. Although it was realized, well before CNT was built, that such a boundary would be beneficial for confinement [17], it was not clear at that time that a large amount of electrons would be on unconfined orbits unless such a boundary was present. A segmented copper mesh, conforming to the outer magnetic surface shape was recently installed and aligned [18]. First experiments with this boundary installed indicate a large improvement in confinement, of nearly a factor of ten. The best confinement times in CNT before mesh installation were on the order of 20 msec, and recently, after mesh installation, the confinement time has been improved to 190 msec. Although it would appear that the particle orbits are vastly improved, this is not necessarily the case. One should keep in mind that there are two dominant sources of transport in CNT: rod-driven, and neutral-driven transport. The previous confinement record of 20 msec was achieved with a neutral pressure of 5×10^{-9} Torr, where rod driven transport was dominant [12], and the new record confinement time was achieved at an even better neutral pressure of 1.7×10^{-9}, where rod-driven transport is also expected to dominate - especially if the transport due to neutral collisions is reduced due to the improved orbit quality. Indeed, the new confinement record is due to a roughly three fold decrease in rod driven transport at a given magnetic field, and due to a 50% increase in the magnetic field strength, ($B = 0.15$ Tesla compared to $B = 0.1$ Tesla previously) which, as mentioned earlier, gives a 50% reduction of rod-driven transport.

DISCUSSION AND CONCLUSION

Confinement in CNT depends primarily on two separate, experimentally distinguishable, radial transport processes: Rod-driven transport, and neutral-collision driven transport.

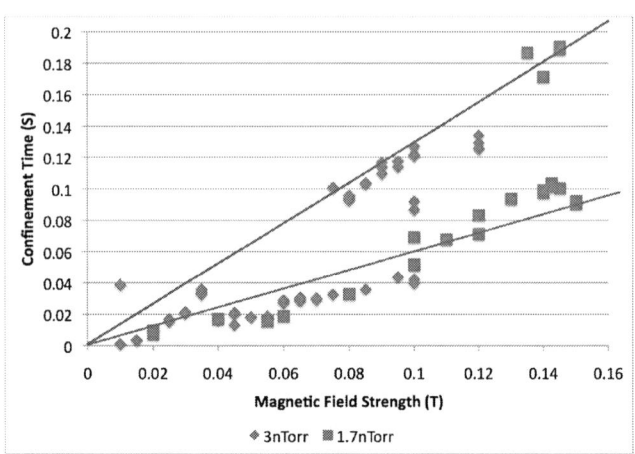

FIGURE 2. With the new mesh installed, the confinement has been improved by an order of magnitude. The longest confinement time, 190 msec, is found at the largest magnetic field strength, $B = 0.15T$. The two straight lines are shown to guide the eye to the two confinement states that are responsible for the vertical scatter in the data.

The former is rather well understood as being due to the $E \times B$ convection cell set up by the electrostatic perturbation of the rod. A model was developed that characterizes the transport rate quite well under varying experimental conditions. This model predicts that the loss rate due to the rod scales as $\Gamma \: n_e^{1/2} T_e^{3/2}$, so confinement, when rod driven transport dominates, should scale as $n_e^{-1/2}$. Comparing the record confinement time experiment with the mesh to the previous confinement time record experiment without the mesh, the magnetic field is 1.5 times larger, the electron temperature is the same, and the electron density is 3 times larger in the most recent case with the mesh. Thus, the total expected confinement improvement should be a factor of $1.5 * \sqrt{3} \approx 3$ rather than ≈ 10 if rod driven transport is dominant and our rod model is correct. It is clear, however, that the model tends to underestimate the actual transport rate, especially for the previous boundary condition in CNT. It also fails to predict the magnitude of the confinement improvement with the flux surface conforming electrostatic mesh installed. We speculate that both of these quantitative discrepancies are due to the simplified model of the Debye screening of the rod, and that the rod driven transport rate is more sensitive to the magnitude of α_D than the model predicts. There is an effort underway to use the 3-D Poisson-Boltzmann equilibrium solver [19] to determine the size of the Debye sheath in CNT. The experimentally measured neutral driven transport shows signs of unconfined orbits for our previous boundary condition. This is confirmed by direct numerical simulations, which also predict vast improvement in the orbit quality with the new mesh installed. Nonetheless, the combination of higher B-field, significantly lower neutral pressure, somewhat lower sensitivity to neutral collisions, and a reduced rod driven transport rate, has led to an order of magnitude improvement in the confinement time, from 20 msec reported earlier [12], to 190 msec [20], see Figure 2. The different

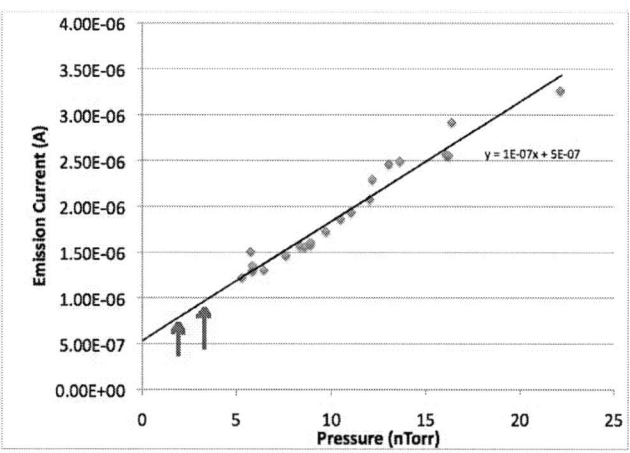

FIGURE 3. A scan of electron loss rate versus neutral pressure in the low neutral pressure range was performed after mesh installation at $B = 0.055$ T. An approximately linear relationship is still seen, with the offset attributed to rod driven transport. The sensitivity to neutral collisions is modestly reduced, by approximately 30%, compared to pressure scan results before the mesh was installed. The two neutral pressures used in Fig. 2 are indicated with red arrows.

data points approximately lie on the two straight lines shown, indicating that confinement improves linearly with B, and that there are two confinement states in CNT [21]. Data from two different (both low) neutral pressures are shown, showing no systematic difference between the two. This, combined with the linear scaling with B, suggests that the neutral driven transport is small compared to the rod driven transport for the discharges being shown in the Figure. Experimentally, the neutral driven transport has not been fully characterized with the mesh installed, but initial results show a rather modest reduction in neutral driven transport at a given neutral pressure - on the order of 30%. A pressure scan was performed, Fig. 3. This shows that indeed the rod driven transport is strongly reduced, and at the two neutral pressures used in Figure 2, $p_n = 1.7$ and 3.0 nTorr, the rod driven transport does dominate over the neutral driven transport, although not by much, since it is vastly reduced relative to its magnitude before mesh installation. This newly achieved long confinement time is an order of magnitude longer than the retraction time of the retractable emitter; thus, experiments with relatively long confinement times without internal objects are now possible - although it still remains to be shown experimentally that the plasmas indeed persist after retraction. This has not yet been achieved in CNT because measurements with capacitive probes of the slow decay of the plasma space charge has so far proven to be challenging. Nonetheless, assuming that there is indeed a substantial amount of plasma left after retraction, it appears that CNT can now produce adequate targets for injection of positrons and creation of electron-positron plasmas. This would require an instantaneous source of approximately 10^{11} positrons, still outside the experimental capabilities of positron traps, but may soon be available [9].

ACKNOWLEDGMENTS

This work was supported by the NSF-DOE partnership in basic plasmas physics and the NSF CAREER program through grants NSF-PHY-04-49813 and NSF-PHY-06-13662, the ORISE Fusion Postdoctoral Fellowship program, and Columbia University.

REFERENCES

1. T. S. Pedersen, A. Boozer, J. Kremer, R. Lefrancois, F. Dahlgren, N. Pomphrey, W. Reiersen, and W. Dorland, *Fusion Science and Technology* **46**, 200 (2004).
2. T. S. Pedersen, J. Kremer, R. Lefrancois, Q. Marksteiner, N. Pomphrey, W. Reiersen, F. Dahlgren, and X. Sarasola, *Fusion Science and Technology* **50**, 372 (2006).
3. J. S. deGrassie, and J. H. Malmberg, *Phys. Rev. Letters* **39**, 1077 (1977).
4. M. R. Stoneking, P. Fontana, and R. Sampson, *Phys. Plasmas* **9**, 766 (2002).
5. A. H. Boozer, *Physics of Plasmas* **11**, 4709 (2004).
6. T. M. O'Neil, and R. A. Smith, *Phys. Fluids B* **4**, 2720 (1992).
7. T. S. Pedersen, and A. H. Boozer, *Phys. Rev. Letters* **88**, 205002 (2002).
8. T. S. Pedersen, A. H. Boozer, W. Dorland, J. P. Kremer, and R. Schmitt, *J. Phys. B* **36**, 1029 (2003).
9. R. G. Greaves, and C. M. Surko, *Radiation Physics and Chemistry* **68**, 419 (2003).
10. J. W. Berkery, T. S. Pedersen, and L. Sampedro, *Review of Scientific Instruments* **78**, 013504 (2007).
11. J. P. Kremer, T. S. Pedersen, et al., *Review of Scientific Instruments* **78**, 013503 (2007).
12. J. P. Kremer, T. S. Pedersen, R. L. Lefrancois, and Q. Marksteiner, *Phys. Rev. Letters* **97**, 095003 (2006).
13. J. W. Berkery, T. S. Pedersen, J. P. Kremer, Q. R.Marksteiner, , R. L. Lefrancois, M. S. Hahn, and P. W. Brenner, *Phys. Plasmas* **14**, 062503 (2007).
14. A. Kabantsev, J. H. Yu, R. B. Lynch, and C. F. Driscoll, *Physics of Plasmas* **10**, 1628 (2003).
15. J. Marler, and M. R. Stoneking, *Phys. Rev. Letters* **100**, 155001 (2008).
16. B. D. de Gevigney et al., *This conference* (2008).
17. T. S. Pedersen, *Phys. Plasmas* **10**, 334 (2003).
18. P. W. Brenner, T. S. Pedersen, J. W. Berkery, Q. R. Marksteiner, and M. S. Hahn, *IEEE Transactions on Plasma Science* **36**, xxx (2008).
19. R. G. Lefrancois, T. S. Pedersen, A. H. Boozer, and J. P. Kremer, *Physics of Plasmas* **12**, 072105 (2005).
20. P. W. Brenner, et al., *AIP Conference Proceedings (this conference)* (2008).
21. M. S. Hahn, et al., *AIP Conference Proceedings (this conference)* (2008).

Studies of a Parallel Force Balance Breaking Instability in a Stellarator

Q.R. Marksteiner[a], T. Sunn Pedersen, J.W. Berkery, M.S. Hahn, J.M. Mendez, B. Durand de Gevigney, P. Ennever, D. Boyle, M. Shulman[b] and H. Himura[c]

[a] *Los Alamos National Laboratory, ISR-6, Los Alamos, NM 87545*
[b] *Department of Applied Physics and Applied Mathematics, Columbia University, New York, New York, 10027, USA*
[c] *Kyoto Institute of Technology, Department of Electronics, Matsugasaki, Kyoto 606-8585*

Abstract. An instability has been observed in non-neutral plasmas confined on magnetic surfaces in the presence of a finite ion fraction [Phys. Rev. Letters **100**, 065002 (2008)]. The dependence of the frequency and amplitude of the instability on neutral pressure, magnetic field strength, and ion species show that the mode consists of interacting perturbations of ions and electrons. In the Columbia Non-neutral Torus (CNT) the instability has a poloidal mode number of m=1. This does not correspond to a rational surface, implying that the parallel force balance of the electron fluid is broken. Here the diagnostic methods used to study this instability are described in detail, and some key results are shown.

Keywords: Non-neutral plasmas, pure-electron plasmas, instability, magnetic surfaces, stellarator, trapped particles.
PACS: 52.27.Jt, 52.35.-g, 52.55.Hc.

INTRODUCTION

CNT is a simple stellarator configuration, consisting of nested magnetic surfaces generated entirely from four circular, planar magnetic coils [1, 2]. The nested magnetic surfaces of a stellarator have a finite magnetic winding number ι, defined as the number of poloidal transits of the magnetic field per toroidal transit. In CNT ι ranges from $\iota = 0.12$ at the magnetic axis to $\iota = 0.23$ at the boundary of the plasma. The underlying physics governing the stability of non-neutral plasmas confined on a magnetic surface configuration like CNT is different from the physics of non-neutral plasmas confined in Penning and Pure-toroidal traps [3, 4]. This is because electron plasmas on magnetic surfaces are in a minimum energy state [5, 6], where low frequency oscillations can in principle damp out by electrons streaming along the magnetic field from a high density part of the perturbation to a low density part. This was thought to happen unless the mode structure of the instability corresponded to a surface where ι is a rational number, and hence electrons could not stream along the magnetic field to damp out the instability.

Despite this, an ion-driven instability has been observed in CNT which has a poloidal mode number of m=1 [7], which does not correspond to a rational surface of CNT. The measured frequency and amplitude dependence of the mode show that the mode consists of interacting perturbations of ions and electrons [7]. The fact that there is an unstable electron component that does not resonate with a rational surface implies that the parallel force balance of the electron fluid is being broken.

The ion driven instability in CNT was diagnosed using internal floating emissive probes and external capacitive probes. This paper describes the diagnostic methods in detail and elaborates on some of the results presented in ref. [7].

PARAMETER DEPENDENCE OF INSTABILITY

Floating emissive probes where used to study how the frequency and amplitude of the instability varied with magnetic field, electric field, neutral pressure, and the ion species introduced. These experiments showed that the unstable mode appears when the ion density exceeds approximately 10% of the ion density. Below this threshold, the plasma is stable. It was also found that the frequency decreased with increasing magnetic field, but did not scale as 1/B, and depended on the species of ion in the plasma. These, and other important results, are shown in ref. [7].

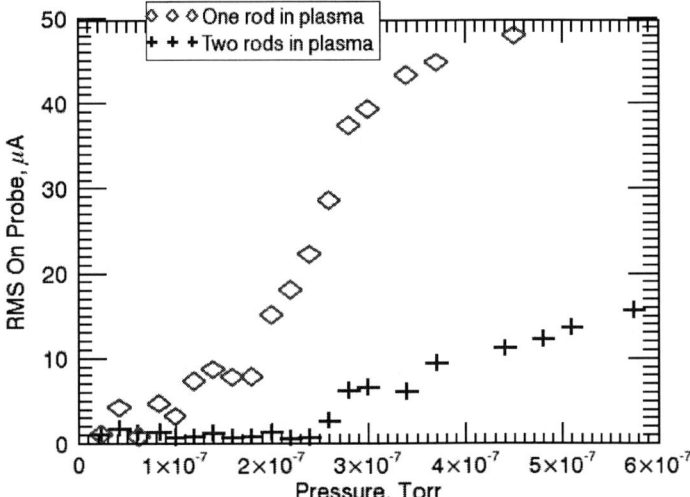

FIGURE 1. Measurements of the RMS of the signal on an emissive probe as a function of neutral pressure of air, with one or two ceramic rods inside the plasma, at B=0.02 T, φ_{plasma}=-200 volts.

Ions are introduced into the plasma by raising the neutral pressure. Electron impact ionization then creates ions in the plasma. The rods which are used to hold the tungsten filaments in the plasma act as a sink for ions, causing the amount of ions to reach a steady state [11].

It was important to determine if it was ions or neutrals that were destabilizing the plasma. This was accomplished by measuring the threshold neutral pressure for instability with one and two ceramic rods inside the plasma. The ceramic rods used for this experiment are at symmetric locations in the plasma (shown in fig. 1 of ref. 9), so that the steady state ion fraction when there are two rods in the plasma will be half of what it is when there is just one rod in the plasma, at a given neutral pressure.

Fig. 1 shows the RMS of the amplitude of the oscillation measured on a floating emissive probe, vs. neutral pressure, with one and two rods in the plasma. The floating emissive probe is 7.6 cm away from the magnetic axis, in the thin cross section of the plasma where the electron density is high. In fig. 1, a large jump in the amplitude of the oscillation is seen at 1.2×10^{-7} Torr when there is one rod in the plasma, and a jump is seen at 2.8×10^{-7} Torr when there are two rods in the plasma. This shows that the threshold for instability is at a given ion fraction, and not at a given neutral pressure. Thus it is ions and not neutrals that are destabilizing the plasma.

CNT has a complicated geometry, where the strengths of the electric and magnetic fields vary by a large amount inside the confining region [9, 1]. The strength of the magnetic field can be adjusted while keeping the magnetic geometry the same by proportionally changing the current in all four magnetic coils [1]. The scalar strength of the magnetic field in CNT is defined as the strength of the magnetic field on the magnetic axis in the thin cross section of the plasma. The electric fields in CNT can be adjusted (approximately proportionally) by varying φ_{plasma}, the potential on the electron emitter. For the experiments shown here, the electric emitter was on axis in the thin cross section of the plasma.

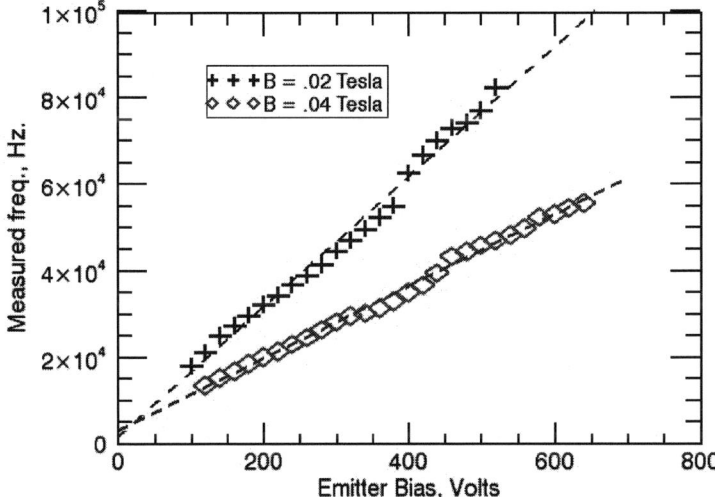

FIGURE 2. The frequency of the instability vs. φ_{plasma} at two different magnetic field strengths. The background neutral pressure is 2.5×10^{-7} Torr of N_2.

Fig. 2 shows the frequency of the mode vs. φ_{plasma}, at two different magnetic field strengths. The mode is increasing approximately linearly with φ_{plasma}, as expected for a mode that scales with the **E**×**B** drift of the electron plasma. The frequency of the mode is decreasing with magnetic field strength. However, the frequency of the mode does not scale exactly as 1/B, and depends on the species of ion introduced [7]. The frequency scaling indicates that the mode involves interacting perturbations of ions and electrons.

Description of Floating Emissive Probe Diagnostic

The measurements described above were made with the same tungsten filaments that were used to measure the electron temperature, density, and the plasma potential [8] in CNT. When being used to measure oscillations, a 10 nF capacitor is placed in between the diagnostic circuit and the probe, to prevent DC current from flowing into the circuit and to allow the heated filament to charge up to close to the plasma potential. It takes approximately 0.5 seconds for the electron current collected by the filament to saturate the voltage of the 10 nF capacitor. Because of this, the plasma is allowed to run for at least 5.5 seconds before data is taken. An op-amp current to voltage converter is used to measure current fluctuations at the probe.

The floating emissive probe has both resistive and capacitive coupling to plasma oscillations. The resistive coupling exists because the hot filament collects and emits electrons as the relation between the local plasma potential and the probe potential changes. The capacitive coupling comes about from image charges in the conducting filament coupling to local potential oscillations. The measured signal from the capacitive coupling is 90° out of phase with the measured signal from resistive coupling. The capacitive coupling of these probes is small ($\sim$ 1 pF) but can compete with the resistive coupling in regions of low density. This phase ambiguity makes it difficult to measure the modal structure of the instability with floating emissive probes, and is one of the reasons why external, capacitive probes were used to measure the modal structure of the instability.

SPATIAL STRUCTURE OF THE INSTABILITY

A set of 4 external image charge probes where built in order to make non-perturbing, purely capacitive measurements of the spatial structure of the instability. These capacitive probes are made of 5 cm copper discs that are each held in place with an aluminum rod. The aluminum rods are thin enough so that they can be bent and manipulated, but are strong enough so that they will hold the probes in a set location when they are installed. The copper disks are attached with metal screws to a tapped bore in the aluminum rods, and are electronically insulated from the aluminum and from the screws with a pair of ceramic washers. The capacitive probes are soldered to shielded cable which is attached to a UHV electrical feedthrough. When the CNT chamber is not under vacuum, the aluminum rods can be adjusted to set the location of the capacitive probes. A picture of this probe array before installation is shown in fig. 4.1 of ref. [12].

In order to measure the poloidal mode number of the instability, the capacitive probes were placed poloidally around the thick ($\varphi = 90°$) cross section of the plasma, where the plasma is approximately cylindrical. Fig. 3 shows a schematic of the location of the 4 probes, along with the last closed flux surface and the magnetic axis. The angle θ is defined using the magnetic axis as the vertex.

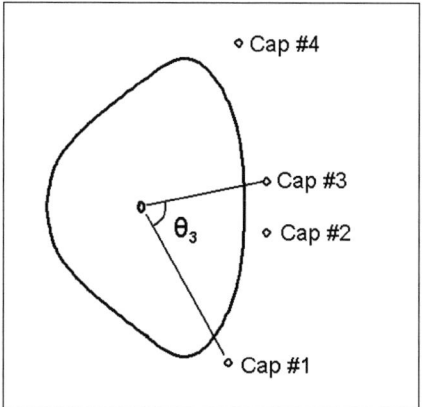

FIGURE 3. The location of the external capacitive probes, along with the last closed flux surface of the $\varphi = 90°$ cross section, and the magnetic axis.

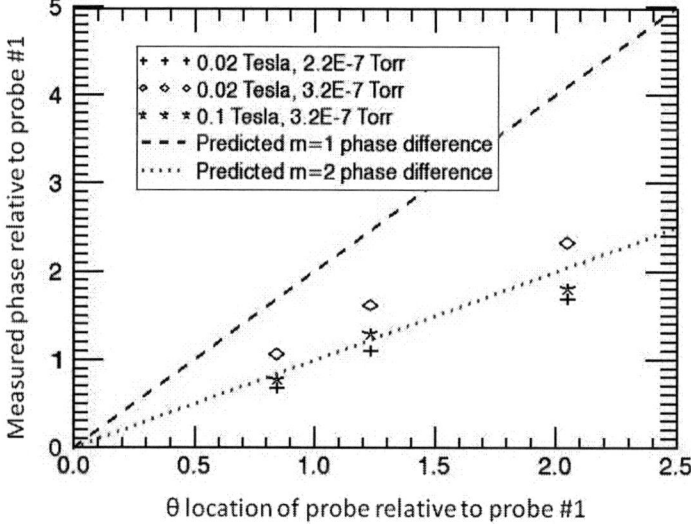

FIGURE 4. The measured phase of the instability on probes 2-4 relative to the phase on probe 1, plotted as a function of the θ difference between these probes and probe 1.

Fig. 4 shows the measured phase between the image charge probes vs. the angle θ between the probes, with probe #1 as a reference, for three different plasma parameters. Also shown are the predicted phase relations if the instability has a poloidal mode number of m=1 or m=2. Although the measured phase difference changes as experimental parameters are varied, the mode is always clearly m=1.

Because there are only a finite number of capacitive probes, it was possible (although unlikely) that there was actually a much higher poloidal mode number, and the apparent m=1 mode was a result of spatial aliasing. In order to verify that this was not the case, the third capacitor was moved by 1 cm, and the experiments were repeated. If spatial aliasing was happening, than the measured phase difference of the third probe would change by a large amount when the probe was moved. This was not observed, confirming that the mode is m=1.

CONCLUSION

In CNT, an instability has been identified that consists of interacting perturbations of ions and electrons. The instability has a poloidal mode number of m=1, which does not correspond to a rational surface. A global instability that does not resonate with a rational surface has never before been observed in a stellarator. Numerical simulations indicate that a large fraction of electrons in CNT are mirror trapped due to parallel variations in the magnetic and electric fields. These trapped electrons cannot stream along the magnetic field to damp out the instability. It is likely that the unstable electron component consists of these trapped particles.

REFERENCES

1. T. S. Pedersen, J.P. Kremer, R.G. Lefrancois, Q. Marksteiner, X. Sarasola, and N. Ahmad, *Phys. Plasmas*, **13**(01):012502, 2006.
2. T. S. Pedersen, J.P. Kremer, R.G. Lefrancois, Q. Marksteiner, N. Pomphrey, W. Reiersen, F. Dahlgren, and X. Sarasola, *Fusion Sci. and Technol.*, **50**:372-381, 2006.
3. J.H. Malmberg and J. S. deGrassie, Phys. Rev. Lett. **35**, 577 (1975).
4. J. D. Daugherty, J. E. Eninger, and G. S. Janes, Phys. Fluids **12**, 2677 (1969).
5. A.H. Boozer, Phys. Plasmas, **11**(10):4709, 2004.
6. T. Sunn Pedersen and A.H. Boozer, Phys. Rev. Lett., **88**(20):205002, 2002.
7. Q.R. Marksteiner, T. Sunn Pedersen, J.W. Berkery, M.S. Hahn, J.M. Mendez, B. Durand de Gevigney, and H. Himura, Phys. Rev. Letters **100**, 065002 (2008).
8. J.P. Kremer, T. Sunn Pedersen, Q. Marksteiner, R.G. Lefrancois, and M. Hahn, *Rev. Sci. Instrum.*, **78**:013503, 2007.
9. J.P. Kremer, T. Sunn Pedersen, R.G. Lefrancois, and Q. Marksteiner, *Phys. Rev. Lett.*, **97**(9):095003, 2006.
10. R.G. Lefrancois, T. Sunn Pedersen, A.H. Boozer, and J.P. Kremer, *Phys. Plasmas.*, **12**(7):072105, 2005.
11. J.W. Berkery, Q.R. Marksteiner, T. Sunn Pedersen, and J.P. Kremer, *Phys. Plasmas.*, **14**:084505, 2007.
12. Q. R. Marksteiner, "Studies of Non-Neutral Ion-Electron Plasmas Confined on Magnetic Surfaces", Ph.D. Thesis, Columbia University, 2008.

Numerical studies of transport in the Columbia Non-neutral Torus

B. Durand de Gevigney, T. Sunn Pedersen and A. H. Boozer

Columbia University, New York, NY 10027

Abstract. The confinement of pure electron plasmas in the Columbia Non-neutral Torus (CNT) stellarator is limited by the presence of unconfined orbits. The existence of a very large electric field across magnetic surfaces should preclude such unconfined orbits. However variations in the electric potential on magnetic surfaces, inherent to the CNT equilibrium, add to the complexity of the trajectories and lead to bad orbits. We have written a code using magnetic coordinates to integrate the electron drift trajectories in the electric and magnetic fields expected in CNT equilibria. Results of such calculations are presented showing that there exists unconfined orbits in CNT if the potential is not constant on surfaces.

Keywords: Non-neutral plasmas, pure electron plasmas, plasma confinement, stellarators, magnetic confinement, neoclassical transport, electron-positron plasmas
PACS: 52.27.Jt,52.27.Ep,52.27.Aj

INTRODUCTION

CNT is a simple stellarator made of only four planar coils [1, 2] and dedicated to the study of plasmas of arbitrary neutrality. This is not an optimized stellarator and it exhibits large variations in the magnitude of the magnetic field on surfaces. As a consequence there exists a large population of trapped electrons. In the absence of an electric field these trapped particles drift out of the plasma very quickly. But because the plasma is non-neutral, a very large electric field develops across the plasma thus closing the orbits by $\vec{E} \times \vec{B}$ rotation. However, inherent to the CNT equilibrium [3, 4], the electric potential is in general not constant on surfaces. This non-constancy of the potential leads to radial $\vec{E} \times \vec{B}$ drift and bad orbits.

ELECTRON ORBITS

Boozer coordinates are known to be useful for studying transport in plasmas confined on magnetic surfaces [5]. These coordinates transform an arbitrarily shaped torus into a standard torus described by radial coordinate ψ, poloidal angle θ and toroidal angle φ. In CNT, neglecting the plasma current because of the very low plasma density [3], these coordinates are related to the magnetic field by

$$\vec{B} = \vec{\nabla}\psi \times \vec{\nabla}\theta + \iota(\psi)\vec{\nabla}\varphi \times \vec{\nabla}\psi = \frac{\mu_0 G}{2\pi}\vec{\nabla}\varphi \quad (1)$$

with $2\pi\psi$ the toroidal flux of $\vec{B}$, $\iota(\psi)$ the rotational transform (the twist of the field lines) and G the total current in the interlocking coils. These coordinates can be calculated by simple field line integration as described in Ref. [6].

Introducing $\theta_0 = \theta - \iota\varphi$, $\chi = \mu_0 G/2\pi$, $\rho_\parallel = mv_\parallel/eB$ - 'parallel Larmor radius' - and $\mu = mv_\perp^2/2B$ - conserved magnetic moment - the drift equations take the simple form

$$\frac{d\psi}{dt} = -\frac{\partial\Phi}{\partial\theta_0} + \left(\frac{eB}{m}\rho_\parallel^2 + \frac{\mu}{e}\right)\frac{\partial B}{\partial\theta_0} \quad \text{(a)} \quad \text{(b)} \quad \frac{d\theta_0}{dt} = \frac{\partial\Phi}{\partial\psi} - \left(\frac{eB}{m}\rho_\parallel^2 + \frac{\mu}{e}\right)\frac{\partial B}{\partial\psi}$$

$$\frac{d\chi}{dt} = \frac{eB^2}{m}\rho_\parallel \quad \text{(c)} \quad \text{(d)} \quad \frac{d\rho_\parallel}{dt} = \frac{\partial\Phi}{\partial\chi} - \left(\frac{eB}{m}\rho_\parallel^2 + \frac{\mu}{e}\right)\frac{\partial B}{\partial\chi} \quad (2)$$

Given an intial pitch $\lambda = v_\parallel/v$ and kinetic energy W_k these equations are integrated using a fourh order accurate Runge-Kutta method with adaptive time step.

In the above equations Φ is the electrostatic potential created by the space charge. Contrary to quasi-neutral plasmas, in CNT Φ can vary on surfaces. Also $e\Delta\Phi \gg T_e$ with $\Delta\Phi$ the potential drop across the plasma and T_e the electron temperature. $T_e \simeq 4eV$ in a typical CNT experiment.

The potential in CNT is derived by solving a Poisson-Boltzmann equation [3, 4]:

$$\nabla^2\Phi = \frac{e}{\varepsilon_0}N(\psi)\exp\left(\frac{e\Phi}{T_e(\psi)}\right) \quad (3)$$

This potential is very sensitive to boundary conditions and is in general not constant on magnetic surfaces.

In the following sections three different potentials are treated. We first analyze the case where there is no electric potential: $\Phi = \Phi_0 = 0$. This is the 'vacuum case', i.e. corresponds to the motion of electrons when no plasma is present. We next consider a potential perfectly constant on surfaces: $\Phi = \Phi_1(\psi)$. As we have seen this is an 'idealized' potential as, in general, the electrostatic potential is not constant on surfaces. Finally we consider a potential function of all three variables: $\Phi = \Phi_2(\psi, \theta, \varphi)$. This potential is obtained by solving Eqn (3). Φ_1 is calculated by averaging Φ_2 over surfaces.

NO ELECTROSTATIC POTENTIAL - THE VACUUM CASE

As explained before, the magnetic field in CNT is not optimized for confining electrons. Indeed because there are only two interlocking coils surrounding the plasma there are huge variations in B within surfaces. These magnetic wells trap a large fraction of the electrons. And these trapped electrons are lost rapidly due to radial magnetic drifts. This is illustrated in Fig. (1a) where we plot the trajectory of a deeply trapped electron projected in the (ψ, θ)-plane. Because the electron is trapped toroidally it cannot take advantage of the rotational transform to close its orbit. Thus it drifts outward until it leaves confinement. Because all surfaces are subject to such variations in the magnitude of B, all surfaces have a potentially large loss cone.

The typical time for a deeply trapped electron ($\lambda \sim 0$) with 4 eV kinetic energy - typical for CNT - to go from the magnetic axis to the plasma boundary by experiencing $\vec{B} \times \vec{\nabla}B$ drifts is given by

$$\delta t = \frac{e\psi^*}{W_k \left| \frac{\partial \ln B}{\partial \theta_0} \right|} \simeq 0.4 ms \tag{4}$$

where $2\pi\psi^*$ is the magnetic flux through the outermost surface enclosing the plasma. Thus we can expect that all trapped electrons be lost in about 1ms.

To get an idea of the confinement when there is no electrostatic potential, we start 1000 electrons randomly distributed on a surface and in pitch and all with 4eV kinetic energy. We then keep track of the fraction of confined electrons as a function of time. This is done for different surfaces and the results are given in Fig. (1b).

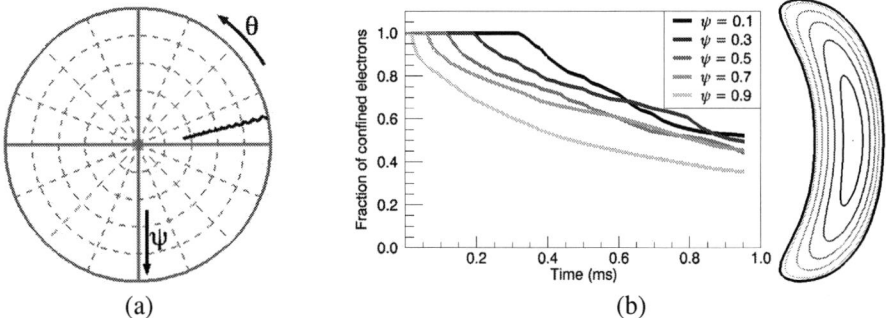

FIGURE 1. (a) Orbit of a deeply trapped electron in Φ_0. The electron does not rotate poloidally and is lost. (b) Fraction of confined electrons for 1000 4eV-electrons started randomly on different magnetic surfaces. On the right are the corresponding surfaces seen on a cross section of the plasma.

ELECTRIC POTENTIAL CONSTANT ON SURFACES - IDEAL CASE

The effect of having an electric potential function of ψ only is to add an $\vec{E} \times \vec{B}$ rotation in the poloidal direction as can be seen from the equations of motion Eqn. (2). This is in addition to the rotation due to the rotational transform $\iota(\psi)$. Nevertheless, for ι to provide poloidal rotation the electron has to travel along the magnetic field, whereas the $\vec{E} \times \vec{B}$ rotation is also effective for trapped particles. This is this additional rotation that saves trapped particles from being lost. This is clear in Fig. (2a) where we plot the trajectory of the same electron as in Fig. (2b) but this time in the potential $\Phi_1(\psi)$.

For this to be of any effect the $\vec{E} \times \vec{B}$ rotation time scale must be short compared with the radial $\vec{B} \times \vec{\nabla} B$ drift time scale. As $e\Delta\Phi \gg T_e$, for thermal electrons

$$\frac{d\Phi}{d\psi} \gg \left(\frac{eB}{m} \rho_\parallel^2 + \frac{\mu}{e} \right) \frac{\partial B}{\partial \psi} = (1+\lambda^2) \frac{W_k}{e} \frac{1}{B} \frac{\partial B}{\partial \psi} \tag{5}$$

where W_k is the kinetic energy of the electron. So that along a trajectory, using Eqn. (2a)-(2b):

$$\frac{d\psi}{d\theta_0} \simeq \left((1+\lambda^2)\frac{\partial \ln B}{\partial \theta_0}\right)\frac{W_k}{e\Phi'} \tag{6}$$

where $\Phi' \equiv d\Phi/d\psi$. The pre-factor is of order unity so that in the limit $e\Delta\Phi \gg T_e$, $d\psi/d\theta_0 \to 0$ and electrons are not allowed to make large excursions from their home surface. So one expects to have bad orbits only in a layer close to the plasma boundary.

One can use Eqn. (6) to estimate the surface ψ_L on which we should begin to observe bad orbits. Trapped electrons are lost if they can make a radial excursion $\Delta\psi$ that can lead them out of the plasma in less time than half a complete poloidal rotation. So there are bad orbits on the surface ψ_0 if:

$$(\Delta\psi)_{max} \equiv \pi \frac{W_k}{e\Phi'}\left(\frac{\partial \ln B}{\partial \theta_0}\right)_{max} > \psi^* - \psi_0 \tag{7}$$

where ψ^* denotes the plasma boundary, the subscript *max* denotes maximum on the surface and all quantities are to be evaluated at $\psi = \psi_0$. Using this estimate for 4eV electrons in the potential Φ_1 gives $\psi_L \simeq 0.85\psi^*$. So apart from a thin layer close to the plasma boundary, the poloidal rotation associated with the electric potential greatly increases confinement. This is confirmed numerically, as illustrated in Fig. (2b).

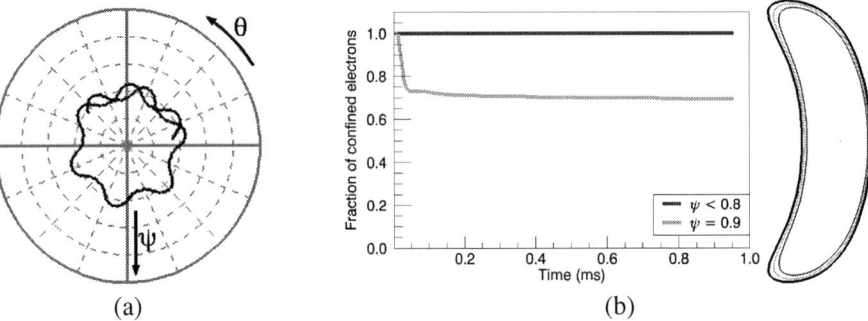

FIGURE 2. (a) Orbit of a deeply trapped electron in Φ_1. The electron rotates poloidally due to $\vec{E} \times \vec{B}$ drift and is well confined. (b) Fraction of confined electrons for 1000 4eV-electrons started randomly on different magnetic surfaces. On the right are the corresponding surfaces.

ELECTRIC POTENTIAL VARYING ON SURFACES

We now turn to the most complicated case of an electric potential depending on all 3 space coordinates. For the first several years of operation in CNT, the electrostatic boundary conditions for solving Eqn (3) were imposed by the interlocking coils and the vacuum chamber - see Fig. (4a) - implying large variations in potential on magnetic surfaces.

The fact that the electric potential can vary on magnetic surfaces adds two new sources for prompt losses. First, variations in Φ_2 in the toroidal direction means that particles can now be trapped in either electric or magnetic wells by the potentials Φ and $\mu B/e$ respectively. These electrostatically trapped electrons add to the magnetically trapped electrons, so the number of electrons possibly subject to prompt losses increases. Second, variations in Φ_2 in the poloidal direction creates radial $\vec{E} \times \vec{B}$ drift. This drift does not depend on the kinetic energy of the electron as do magnetic drifts. This means that even low energy electrons can make significant radial excursions. And by doing so they can pick up energy from the electric field thus increasing their kinetic energy and their magnetic drifts and may transition form trapped to passing.

The combination of all these effects makes analytical calculation impractical. The complexity of the orbits in Φ_2 is illustrated in Fig. (3a) where we plot the trajectory of the same electron as in Fig. (1a) and Fig. (2a) but in the potential Φ_2. However numerical integration of the orbits shows that there is a large fraction of unconfined orbits even deep inside the plasma, see Fig. (3b).

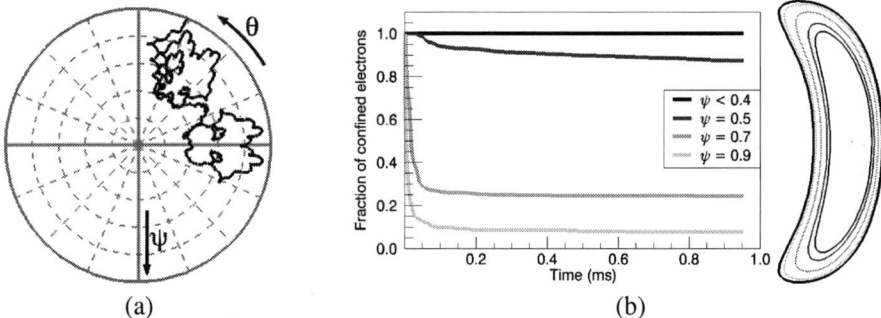

FIGURE 3. (a) Orbit of an electron in Φ_2. The complexity of the orbit is apparent here. (b) Fraction of confined electrons for 1000 4eV-electrons started randomly on different magnetic surfaces. On the right are the corresponding surfaces.

Intuitively this is understood as follows. Without electric field and neglecting magnetic drifts, electrons circulate all around magnetic surfaces following field lines. Now if there is an electric field, electrons also $\vec{E} \times \vec{B}$ drift on equipotential surfaces because $\vec{v}_{\vec{E}\times\vec{B}} = \vec{E} \times \vec{B}/B^2 = -\vec{\nabla}\Phi \times \vec{B}/B^2$ so that $\vec{v}_{\vec{E}\times\vec{B}} \cdot \vec{\nabla}\Phi = 0$. And if the equipotential surfaces do not match magnetic surfaces, electrons can jump from one magnetic surface to the other. After a few steps of jumping from one surface to the other an electron can find its way out of the plasma, see Fig. (4). Simulations show that this process is quite effective and can remove electrons from the plasma in tens of microseconds, see Fig. (3b).

Recently a conducting boundary conforming to the plasma shape was installed, see Fig. (4b). In addition to its use as an external capacitive probe, it reduces the mismatch between electrostatic equipotentials and magnetic surfaces. As expected from above simulations this resulted in an improved confinement time. For more on this see paper by P. W. Brenner at this conference.

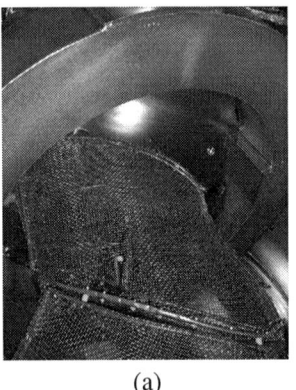 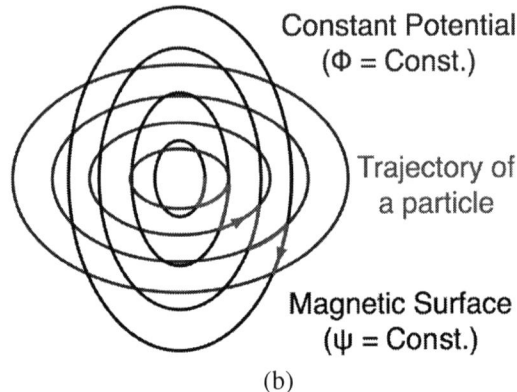

FIGURE 4. (a) Inside CNT. One can see one of the interlocking coils and the recently installed copper mesh. (b) Electrons jump from magnetic surfaces to magnetic surfaces by drifting on equipotentials and can find their way out of the plasma.

CONCLUSION

Numerical investigation of the collisionless orbits of electrons confined in CNT shed some light on measured confinement times well below those estimated in Ref. [7]. It is shown that there exist a large fraction of trapped electrons in CNT. If the potential in CNT were perfectly constant on surfaces the strong induced radial electric field would close the orbits of such trapped particles and would effectively reduce prompt losses. However the potential is not constant on surfaces in the general case and this, in turn, creates bad orbits and certainly explains the poor confinement times measured. The installation of a conducting boundary conforming to the plasma shape improved the constancy of the plasma on magnetic surfaces and did increase confinement time.

ACKNOWLEDGMENTS

This work was supported by the NSF-DOE partnership in basic plasmas physics and the NSF CAREER program through grants NSF-PHY-04-49813 and NSF-PHY-06-13662, the ORISE Fusion Postdoctoral Fellowship program, and Columbia University.

REFERENCES

1. T. S. Pedersen, et al., *Fusion Science and Technology* **46**, 200, 2004.
2. T. S. Pedersen, et al., *Fusion Science and Technology* **50**, 372, 2006.
3. T. S. Pedersen, and A. H. Boozer, *Phys. Rev. Letters* **88**, 205002, 2002.
4. R. G. Lefrancois, et al., *Physics of Plasmas* **12**, 072105, 2005.
5. A. H. Boozer, *Phys. Fluids* **24**, 1999, 1981.
6. G. Kuo-Petravic, et al., *J. Comput. Phys.* **51**, 261, 1983.
7. J. W. Berkery, and A. H. Boozer, *Physics of Plasmas* **14**, 104530, 2007.

Studies Of Enhanced Confinement In The Columbia Non-Neutral Torus

P. W. Brenner*, T. Sunn Pedersen*, M. Hahn*, J. W. Berkery*, R. G. Lefrancois* and Q. R. Marksteiner*,†

*Columbia University, New York, NY 10027
†Los Alamos National Laboratory, Los Alamos, NM 87545

Abstract. Recently the measured confinement time in the Columbia Non-neutral Torus (CNT) has been increased by nearly an order of magnitude to 190 ms. Previously, enhanced transport caused in part by the mismatch of constant potential and magnetic surfaces limited confinement times to 20 ms. A conducting boundary conforming to the last closed magnetic flux surface has been installed to minimize potential variation along magnetic surfaces, provide new methods to influence the plasma, and act as an external diagnostic. A summary of new results with the conducting boundary installed will be presented, including discussion of how confinement is influenced by neutral pressure, magnetic field strength, and the effect of biasing individual sectors of the mesh.

Keywords: Non-neutral plasmas, pure electron plasmas, plasma confinement, stellarators, magnetic confinement, neoclassical transport
PACS: 52.27.Jt,52.27.Ep,52.27.Aj

INTRODUCTION

The Columbia Non-neutral Torus (CNT) is a stellarator designed to study the equilibrium [1], stability [2], and confinement [3] of non-neutral plasmas confined on magnetic surfaces. CNT utilizes a simple four planar coil design to create surfaces so design and construction were relatively quick and inexpensive [4]. With this coil set it is possible to create magnetic surfaces at an aspect ratio of $A \leq 1.9$ without significant internal islands [5]. Steady state pure electron plasmas are created by thermionic emission from a heated and biased tungsten filament held on the magnetic axis by insulating ceramic rods. Parallel transport quickly fills the axis and the remaining surfaces are filled by cross surface transport.

Very long confinement times are predicted for pure electron plasmas confined in a stellarator field [6]. However, detailed experimental confinement studies have been performed [3] and until recently 20 ms was the best confinement time achieved in CNT [1]. Recent improvements including the installation of a conducting boundary conforming to the last closed magnetic flux surface have resulted in a new record confinement time of 190 ms. With the installation of the conducting boundary two different stable confinement states are now measured and it is possible to switch between these states by biasing sectors of the conducting boundary.

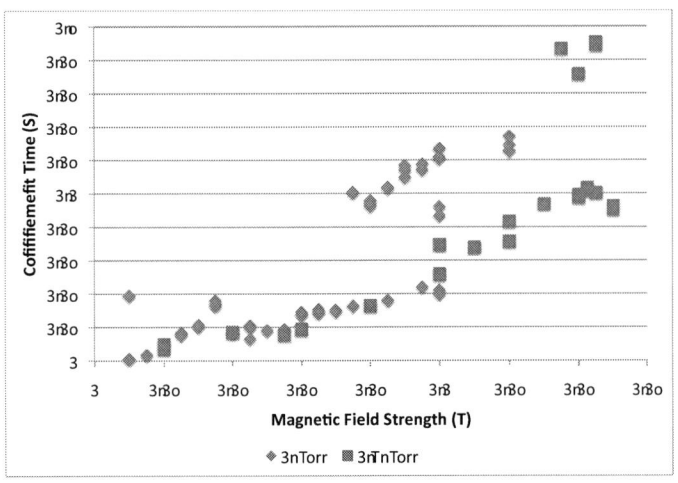

FIGURE 1. Confinement times of 190ms have recently been achieved. Multiple stable confinement states are possible at the same magnetic field. These states are being studied and are discussed later in this paper.

SOURCES OF TRANSPORT

The original 20 ms confinement record was primarily limited by two major sources of transport. The dominant source is a result of the insulating ceramic rods used to hold probes. The rods develop a negative charge causing electrons to **E X B** drift out of the plasma. Also, a mismatch between equipotential surfaces and magnetic surfaces creates a step size approximately equal to the size of the mismatch [7]. Electron neutral collisions, which would only be responsible for relatively slow transport if not for the surface mismatch, then drive a significant portion of the losses. Although confinement without insulating rods perturbing the plasma has not been measured successfully yet, the base pressure and transport due to rods have both been reduced resulting in the new best confinement time. This new record confinement time is visible in the confinement time versus magnetic field plot shown in figure 1 for a 200V plasma at 1.7 nTorr and 3 nTorr base pressures.

Insulating rods in the plasma

Plasmas are created in CNT by emitting electrons from filaments attached to ceramic rods inserted into the magnetic surfaces. Equilibrium and confinement are measured

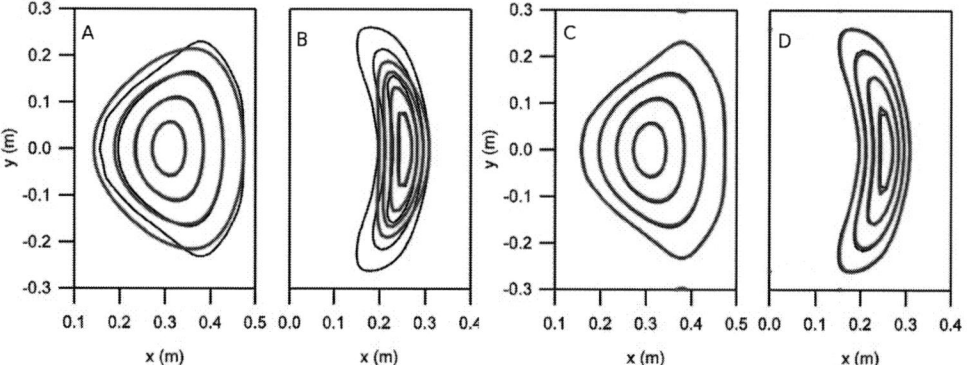

FIGURE 2. Simulations of a cold plasma in CNT show a mismatch between magnetic surfaces (black) and constant potential surfaces (red) when the boundary conditions are set by the interlocking coils and vacuum chamber alone (A and B). Imposing an equipotential at the last closed magnetic surface results in a favorable match (C and D) [8].

with emissive probes attached to the same rods. These insulating ceramic rods charge up negatively causing **E X B** drift along the rod [3]. At current base pressures of 1.7 nTorr, insulating rods account for the majority of transport [9].

A pneumatically operated emitter capable of retracting in 20 msec has been constructed and installed. This retractable emitter can create plasmas and then quickly be retracted to minimize perturbation and transport. With the probing capabilities of the conducting boundary this combination provides a nearly nonintrusive method of creating and diagnosing plasmas and should allow for a proper test of the predicted confinement time scaling [10].

Potential surface deviation from magnetic surfaces

A conducting boundary composed of 13 electrically isolated meshes that conform to the last closed magnetic surface has been installed, creating an equipotential boundary condition. This improves the match between equipotential and magnetic surfaces in the plasma as well as offering new non intrusive methods to diagnose the plasma. Previously potential boundary conditions were set by the shape of the interlocking magnetic coils and the vacuum chamber.

An equilibrium reconstruction code was used [8] to calculate contours of constant potential for cases with and without a conducting boundary that conforms to the last closed magnetic flux surface. Results of the simulation are shown in figure 2. Without a conducting boundary there are large variations of potential contours from magnetic surfaces near the edge driving transport in that region. With a conducting boundary aligned to the last closed flux surface the magnetic and constant potential surfaces are in good agreement throughout the confining region. Improvements in confinement with the installation of the conducting boundary have been observed; however, the reported

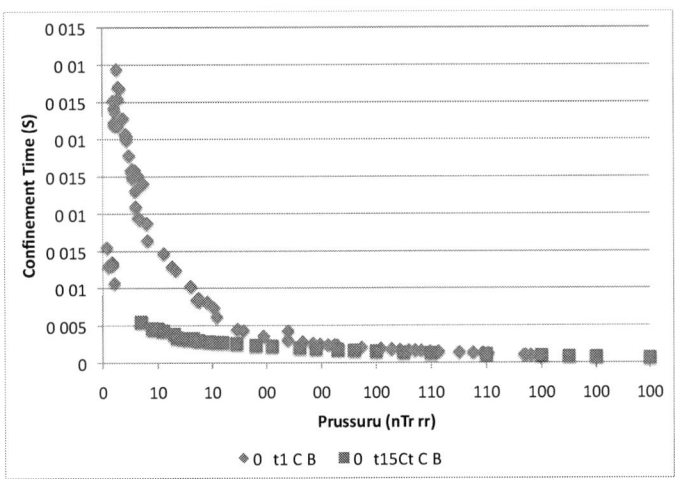

FIGURE 3. Confinement time increases with decreasing pressure as expected. Longer confinement times are seen with the conducting boundary installed. There is also a state at low pressure with relatively poor confinement.

order of magnitude improvement includes improvements due to a higher magnetic field, lower pressure, and a decrease in rod driven transport [9].

Neutral driven transport

Collisions with neutrals drive an increasing amount of transport as pressure is increased. By measuring the electron loss rate at various pressures and then extrapolating to zero pressure it is possible to estimate transport caused by sources other than electron neutral collisions, presumably the insulating rods. At the current base pressure of 1.7 nTorr, pressure driven losses are less than the sum of the other sources of transport. An evaporable lithium pump is currently being investigated to further lower the pressure to a point where neutral driven transport losses would be insignificant compared to rod driven transport. Then using the retractable emitter described in the previous section it should be possible to study much smaller sources of transport and achieve significantly longer confinement times.

A plot of confinement time versus pressure for 200V plasmas with a magnetic field of 0.055T is shown in figure 3. Two stable confinement states are again obvious in the data shown.

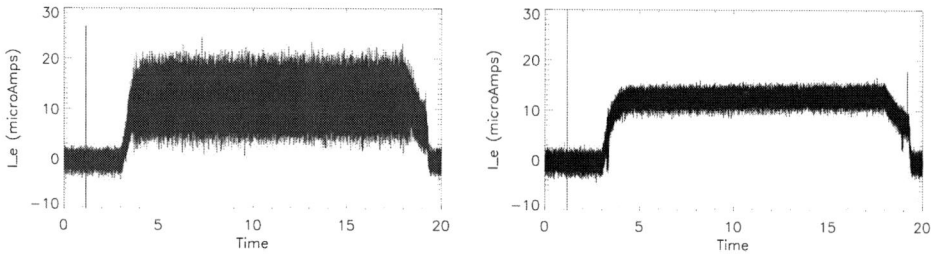

FIGURE 4. Two plasmas with approximately the same confinement time at different pressures are shown. In red on the left is a plasma in the low confinement state at 25 nTorr. In blue on the right is a plasma in the high confinement state (same emission current but higher pressure) at 46 ntorr.

CONFINEMENT STATES

Since the installation of the conducting boundary two stable states with different confinement times are observable for certain conditions. Thus it is possible to decrease the pressure but measure a worse confinement time if the plasma switches from a high to a low confinement state at the lower pressure. This is clear in figure 3 where the confinement time for pressures below 5 nTorr are approximately the same as confinement times around 20 nTorr. Figure 4 shows the emission current (proportional to the loss rate) for 2 shots at different pressures. Although the pressure for the red trace is nearly half of pressure for the blue trace they both have approximately the same average emission current. It is also clear in the figure that the low confinement state exhibits much larger oscillations in the emission current than the high confinement states. Further measurements will be done to learn how else the plasma equilibrium differs between the two states and to study the oscillations seen.

Currently these confinement states are being investigate to discover if they are the same as the confinement jumps that are being studied at higher loss rates [11]. Although these confinement states are not yet fully understood it is possible to provoke the plasma from low confinement to the high confinement state. This is done by pulsing a -100V bias on a segment of the conducting boundary near the emission filament. During the pulse the emission current drops and after the pulse the plasma remains in the high confinement state.

CONCLUSION

Progress is being made in the goal to approach the very long confinement times predicted possible for CNT. Recently, with a higher magnetic field, lower base pressure, and the installation of a conducting boundary conforming to the last closed magnetic flux surface a new record confinement time of 190 ms has been achieved. Further work is being done to use sections of the conducting boundary as external diagnostics. Combined with the use of a retractable emitter and decreased base pressures it should be possible to measure confinement that is no longer limited by rod and neutral driven transport. The ability to provoke the plasma into different stable confinement states provides an interesting new tool to study the equilibrium in CNT. Further work is planned to study how the plasma can be influenced by the conducting boundary and understand these different confinement regimes.

ACKNOWLEDGMENTS

This work was supported by the NSF-DOE partnership in basic plasmas physics and the NSF CAREER program through grants NSF-PHY-04-49813 and NSF-PHY-06-13662, the ORISE Fusion Postdoctoral Fellowship program, and Columbia University.

REFERENCES

1. J. Kremer, T. S. Pedersen, Q. Marksteiner, and R. Lefrancois, *Physical Review Letters* **97** (2006).
2. Q. Marksteiner, T. S. Pedersen, J. Berkery, M. Hahn, J. Mendez, B. D. de Gevigney, and H. Himura, *Physical Review Letters* **100**, 065002 (200).
3. J. Berkery, T. S. Pedersen, J. Kremer, Q. Marksteiner, R. Lefrancois, M. Hahn, and P. W. Brenner, *Physics of Plasmas* **14** (2007).
4. T. S. Pedersen, A. Boozer, J. Kremer, R. Lefrancois, W. Reiersen, F. Dahlgreen, and N. Pomphrey, *Fusion Science and Technology* **46**, 200–208 (2004).
5. T. S. Pedersen, J. Kremer, R. Lefrancois, Q. Marksteiner, N. Pomphrey, W. Reiersen, F. Dahlgreen, and X. Sarasola, *Fusion Science and Technology* **50**, 372–381 (2006).
6. T. S. Pedersen, and A. Boozer, *Physical Review Letters* **88**, 205002 (2002).
7. J. Berkery, Q. Marksteiner, T. S. Pedersen, J. Kremer, and R. Lefrancois, *Physics of Plasmas* **14**, 084505 (2007).
8. R. Lefrancois, T. S. Pedersen, A. Boozer, and J. Kremer, *Physics of Plasmas* **12**, 072105 (2005).
9. T. S. Pedersen, J. W. Berkery, and et al., *AIP Conference Proceedings (this conference)* (2008).
10. J. W. Berkery, and A. H. Boozer, *Phys. Plasmas* **14**, 104503 (2007).
11. M. Hahn, T. S. Pedersen, and et al., *AIP Conference Proceedings (this conference)* (2008).

Pure Electron Equilibrium and Transport Jumps in the Columbia Non-neutral Torus

M. Hahn*, T. Sunn Pedersen*, J. W. Berkery*, Q. R. Marksteiner*,[†], P. W. Brenner* and B. Durand de Gevigney*

*Columbia University, New York, NY 10027
[†]Los Alamos National Laboratory, Los Alamos, NM 87545

Abstract. The Columbia Non-neutral Torus (CNT) is a simple stellarator being used to study non-neutral plasmas. At low neutral pressures these plasmas are pure electron plasmas. In this paper some studies of the equilibrium and transport of these plasmas will be described. An axial density variation has now been measured that is roughly the size predicted numerically. The equilibrium of plasmas created by an off-axis emitter have properties that are consistent with global thermal equilibrium. Sudden jumps in emission current have been observed that correspond to jumps in transport between two stable equilibrium states.

Keywords: Non-neutral plasmas, pure electron plasmas, plasma confinement, stellarators, magnetic confinement
PACS: 52.27.Jt,52.27.Ep,52.27.Aj

INTRODUCTION

The Columbia Non-neutral Torus is a simple stellarator [1, 2] being used to study non-neutral plasmas. At very low neutral pressures these plasmas are pure electron plasmas. The equilibrium of such plasmas is determined by electrostatic force balance [3]. One consequence of this type of equilibrium is that the geometry of the magnetic surfaces plays a significant role. In CNT the geometry causes a density and potential variation along the magnetic axis. Another property of CNT equilibria is that when emitting from off the magnetic axis the plasma inside the surface of the emitter has an equilibrium consistent with thermal equilibrium across the magnetic surfaces.

Transport in CNT plasmas is determined in part by the coupling of the emission current to the loss rate through the continuity equation, and by the plasma equilibrium. Jumps are observed in the emission current at certain transport rates along with changes in the plasma equilibrium. Experimental studies strongly suggest that the jumps are caused by a cathode instability.

The experimental studies described here use measurements from internal particle flux probes. Plasma potential is measured using both emissive (hot) and Langmuir (cold) probes to identify the deviation potential. Temperature and density are measured using the Langmuir probe current-voltage characteristic. Detailed descriptions of the experimental methods can be found in Ref. [4]. Numerical results were obtained by solving the Poisson-Boltzmann equation in the magnetic surface geometry of CNT using the method described in Ref. [5].

FIGURE 1. Three dimensional view of the magnetic surfaces in CNT.

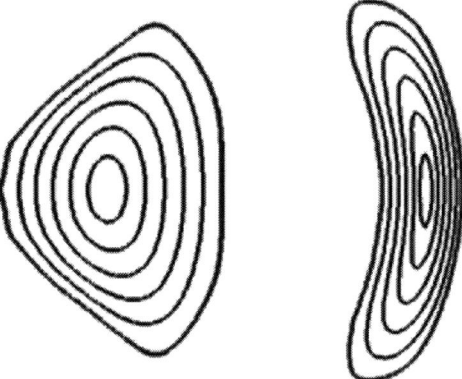

FIGURE 2. Magnetic surface cross sections at toroidal angle $\phi = 90°$(left) and $\phi = 0°$ (right)

AXIAL DENSITY VARIATION

CNT has a large toroidal variation in the shape of the magnetic surfaces (Fig. 1). In particular, there are significant differences in the cross-sectional shape of the magnetic surfaces between the $\phi = 0°$ and $\phi = 90°$ locations (Fig. 2). This toroidal variation in the shape of the magnetic surfaces leads to a density variation along the magnetic axis [6, 7].

In the cold plasma limit the radial potential drop from the axis to the edge at each cross-section must be the same. Poissons equation implies that if the cross-sectional shape is varying there must be a corresponding variation of the density. Typical CNT plasmas have a small Debye length (≈ 1.5 cm), so they are close to this limit and should have the density variation. CNT plasmas also have a finite temperature with electrons that follow a Boltzmann distribution. Since the Boltzmann distribution relates the density to the potential the density variation will be accompanied by a potential variation.

A numerical code was used to study the equilibrium in the magnetic surface geometry of CNT and predict the size of the density variation. For a boundary condition approximating the shape of the vacuum chamber and magnetic field coils the variation was predicted to be $n_0/n_{90} = 4.5$ and $e\Delta\Phi/T_e = 1.5$. For a conducting boundary whose shape conforms to the outermost magnetic surface the variation was slightly larger with $n_0/n_{90} = 5.3$ and $e\Delta\Phi/T_e = 1.7$ [6].

The density and plasma potential at the two toroidal locations was measured for several emitter biases and the results were averaged to arrive at a median value and uncertainty range [7]. The density variation was found to be $n_0/n_{90} = 7.8$ with a 1σ uncertainty range between 3.8 and 20.0. The potential variation was measured to be $e\Delta\Phi/T_e = 1.1$ with a 1σ uncertainty range of 0.7 - 1.3. The measured density variation is in agreement, but with large error bars, while the potential variation is somewhat smaller than predicted. The discrepancy in the potential variation is probably caused by the approximations to the actual boundary conditions used in the numerical solver.

The variations have also been measured with the conforming conducting boundary. Those measurements showed a significantly larger than predicted variation with $e\Delta\Phi/T_e = 4.0$ with a 1σ range 3.0 - 5.5. One possible explanation for this significantly enhanced variation is that electron density is strongly radially peaked. Since the numerical models use density profiles as an input, these unanticipated density profiles were not accounted for in the numerical models. New equilibrium calculations are being done with more accurate density profiles to refine the theoretical predictions.

OFF-AXIS EQUILIBRIUM

For most experiments plasmas are created by emission from a heated biased filament placed on the magnetic axis. Electrons diffuse outward filling the surfaces. Plasmas can also be made by emitting from off the magnetic axis. The equilibrium properties of plasmas from off axis emission were studied for both boundary conditions.

When emitting off-axis the plasma potential becomes more negative on the surfaces inside the location of the emitter by a factor of $\alpha T_e/e$ with α slowly increasing as the distance of the emitter from the magnetic axis increases (Fig. 3). For each boundary condition the electron temperature is constant across the inner surfaces. For the non-conforming boundary condition the magnitude of the constant temperature is $\approx 4\text{eV}$ independent of emitter location. With the conforming boundary the temperature grows with radial emitter location, resulting in larger potential differences than in the non-conforming boundary case. These characteristics are consistent with thermal equilibrium across the inner surfaces.

Numerical reconstructions were done to test the consistency of the off-axis equilibrium with global thermal equilibrium across the inner surfaces. The thermal equilibrium condition can be imposed numerically and the potential profiles output from the code show good agreement with experimental potential measurements (Fig. 4). That this is a non-trivial consistency was shown using other numerical reconstructions to show that

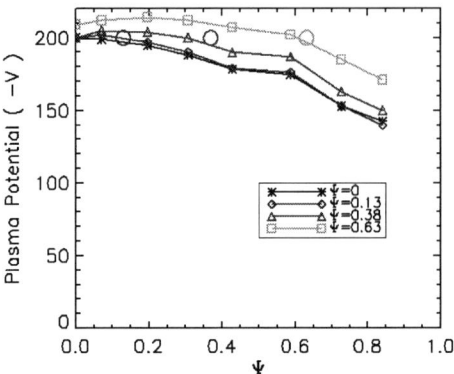

FIGURE 3. Plasma potential measurements for different emitter locations. Ψ is a radial coordinate normalised so that $\Psi = 0$ is the magnetic axis, $\Psi = 1$ is the outermost surface. Circles are centered on the points ($\Psi_{emitter}, V_{emitter}$).

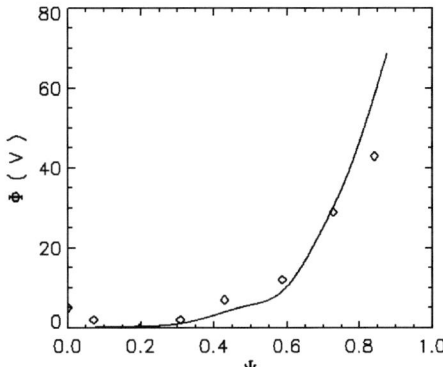

FIGURE 4. Comparison of a numerically calculated potential profile for global thermal equilibrium on the inner surfaces (solid line) with experimental data when emitting from the $\Psi = 0.63$ surface.

imposing cross-surface thermal equilibrium in the code cannot reproduce measurements from on-axis equilibrium, and that the input parameters that model on-axis equilibrium cannot be used to match off-axis equilibrium data.

TRANSPORT JUMPS

Sudden jumps in the emission current are observed as parameters that drive transport are varied. Figure 5 shows jumps in the emission current as the emitter bias is made more

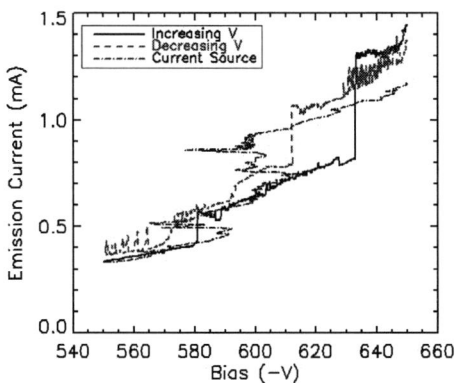

FIGURE 5. Current jumps in the cathode current-voltage characteristic as the emitter bias is made more negative. Hysteresis is observed when the bias is decreased, and the negative differential resistance can be seen by using a current source emitter.

negative. Reducing the emitter bias shows that the jumps are hysteretic. A current source emitter can be used to trace out the I-V characteristic at the jumps to show that the jumps occur over an unstable negative differential resistance.

Parameters are varied slowly enough that the plasma is in steady state so that the emission current equals the loss rate of electrons from the plasma. Jumps in emission current are jumps in transport. Jumps can be caused by varying parameters that drive transport besides emitter bias, for example increasing neutral pressure, or inserting insulating rods while keeping other parameters constant will cause jumps. One interesting observation is that parameters that increase or suppress ionization will both cause jumps as long as the total transport increases. This is in contrast to the situation in quasi-neutral plasmas where hysteretic current jumps depend explicitly on ionization.

Jumps are observed at particular emission currents for any parameter used to drive transport. These currents have not changed with the installation of the conducting boundary, nor do they change with emitter location. Experiments using two emitters show that the jumps occur at particular local emission currents, not the total transport rate from the plasma. If the emitting filaments are aligned so that they occupy the same sheath then the jumps will occur at the total current. These results strongly suggest that the jumps are caused by a cathode instability.

Transport jumps are accompanied by a change in the equilibrium. Plasma potential profiles in the equilibrium states have been measured by using the hysteresis effect to measure the plasma in both the high current and low current states with the same emitter bias (Fig. 6). These potential profile measurements together with numerical results confirm that the change in the total number of electrons confined is small, so the jumps are abrupt drops in the confinement time. At these high transport rates the confinement time is only on the order of 1 ms but this is still much greater than the parallel transport timescale so both states are equilibrium states.

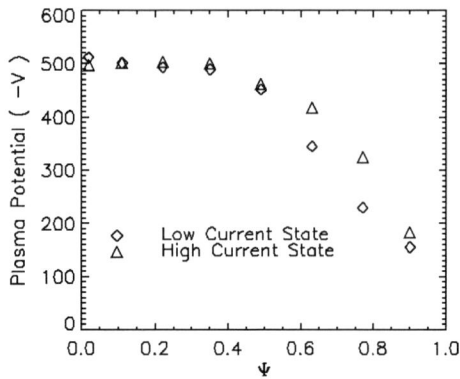

FIGURE 6. The plasma potential profile changes significantly between the high and low current states.

CONCLUSION

Pure electron plasmas confined on magnetic surfaces exhibit many unique properties. In CNT large density and potential variations along the magnetic axis has been measured. Plasmas created by emission from off the magnetic axis have an equilibrium consistent with being in a state of global thermal equilibrium on the inner surfaces. Transport jumps have been observed at particular transport rates. The jumps are probably caused by a cathode instability, which causes a change in the equilibrium to a state that can accomodate a higher transport rate.

ACKNOWLEDGMENTS

This work was supported by the NSF-DOE partnership in basic plasmas physics and the NSF CAREER program through grants NSF-PHY-04-49813 and NSF-PHY-06-13662, the ORISE Fusion Postdoctoral Fellowship program, and Columbia University.

REFERENCES

1. T. S. Pedersen, et al., *Fusion Science and Technology* **46**, 200 (2004).
2. T. S. Pedersen, et al., *Fusion Science and Technology* **50**, 372 (2006).
3. T. S. Pedersen, and A. H. Boozer, *Phys. Rev. Letters* **88**, 205002 (2002).
4. J. P. Kremer, et al., *Rev. Sci. Instrum.* **78**, 013503 (2007).
5. R. G. Lefrancois, et al., *Physics of Plasmas* **12**, 072105 (2005).
6. Remi G. Lefrancois, and Thomas Sunn Pedersen, *Phys. Plasmas* **13**, 120702 (2006).
7. M. Hahn, et al., *Phys. Plasmas* **15**, 020701 (2008).

SECTION III
COLLECTIVE MODES AND TRANSPORT

Electron Acoustic Waves in Pure Ion Plasmas

Francois Anderegg, C. Fred Driscoll, Daniel H.E. Dubin and
Thomas M. O'Neil

Department of Physics, University of California at San Diego, La Jolla, CA USA 92093-0319

Abstract.
Electron Acoustic Waves (EAWs) are the low frequency branch of electrostatic plasma waves. These waves exist in neutralized plasmas, pure electron plasmas and pure ion plasmas. At small amplitude, EAWs have a phase velocity $v_{ph} \simeq 1.4\bar{v}$ and their frequencies are in agreement with theory. At moderate amplitudes, waves can be excited over a broad range of frequencies and their phase velocity is in the range of $1.4\bar{v} \leq v_{ph} \leq 2.1\bar{v}$. This frequency variability comes from the plasma adjusting its velocity distribution so as to make the plasma mode resonant with the drive frequency. These plasma waves can also be excited with a chirped frequency drive resulting in extreme modification of the particle distribution, giving almost undamped waves with $(\gamma/\omega \sim 10^{-5})$.

Keywords: non-neutral plasma, plasma wave, Trivelpiece-Gould wave, EAW
PACS: 52.27.Jt, 52.35.Fp, 52.35.Sb

INTRODUCTION

The near-linear Electron Acoustic Wave (EAW) has a phase velocity $v_{ph} \simeq 1.4\bar{v}$ and its frequency has a strong temperature dependence $f_{EAW} \propto T^{1/2}$. These waves have been studied theoretically [1] and numerically [2]; they have been observed in experiments with pure electron plasmas [3] and in laser produced plasmas [4, 5]. Linear Landau damping suggests that waves with such slow phase velocity are strongly damped. At finite amplitudes, however, trapping of particles near the phase velocity flattens the distribution function, resulting in weakly damped waves.

We observe that at small amplitude, the EAW dispersion relation is correctly described by the Holloway and Dorning approach. In contrast, at larger amplitude, we observe that we can excite EAW-type plasma waves over a continuous range of frequencies encompassing the small amplitude EAW and the small amplitude Trivelpiece-Gould (TG) plasma wave. Under these conditions, these plasma waves can not be uniquely named.

With a chirped frequency drive [6], these waves can also be excited over a continuum of frequencies. These chirp driven waves were initially described as BGK type plasma waves by Fajan's group at Berkeley. Vlasov-Poisson simulations [7] have also investigated the highly non-linear amplitude regime, suggesting that EAW-like modes with strong harmonic content, called KEEN waves, can be excited over a wide range of frequencies.

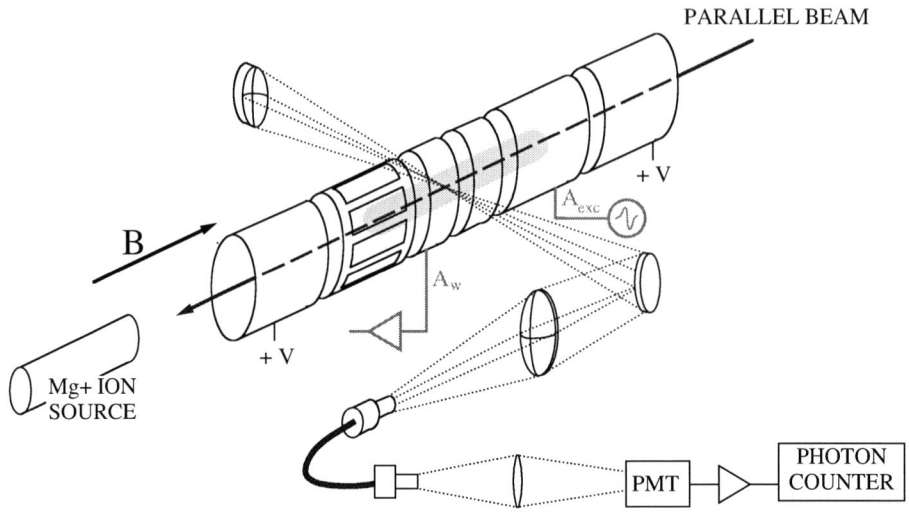

FIGURE 1. Experimental set-up.

APPARATUS

Figure 1 shows the electrode arrangement of our Penning-Malmberg trap with uniform axial magnetic field $B = 3$ Tesla. The trap conducting electrodes have a wall radius $R_w = 2.86$ cm and are contained in ultrahigh vacuum at $P \approx 10^{-10}$ Torr. The ion density is $n \sim 1.5 \times 10^7$ cm^{-3} over a radius $R_p \approx 0.45$ cm with a plasma length $L_p \simeq 9$ cm. The ions are held in steady state for days with a weak "rotating wall" electric field applied to the sectored electrode. The rotating wall is turned off about 100 ms before the wave measurement and turned back on about 200 ms later. The plasma density and temperature are measured with laser induced fluorescence. The temperature is controlled over 0.3 eV $\leq T \leq 1.5$ eV.

We excite standing EAW and TG by applying a 3–100 cycle burst to the end electgrode at a frequency f_{exc} and amplitude A_{exc}. To excite TG waves, 3–10 cycles are generally sufficient; in contrast, EAWs require a much longer drive to excite a wave lasting thousands of cycles. The amplitude of the burst is rounded to avoid exciting spurious TG modes from the harmonic content of the burst while driving an EAW. We excite the longest possible wavelength ($m_z = 1$), that is $\lambda \approx 2L_p$, the lowest radial mode ($m_r = 1$) and only consider azimuthally symmetric modes ($m_\theta = 0$) since excitation and detection electrodes are both azimuthally symmetric. These waves are standing and reflect thousands of times at the plasma ends.

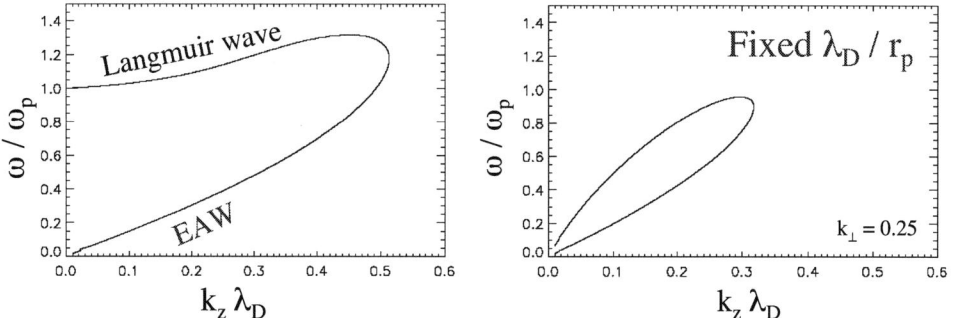

FIGURE 2. (a) Plasma wave dispersion relation in infinite plasma. (b) Plasma wave dispersion in non-neutral finite radial size plasma.

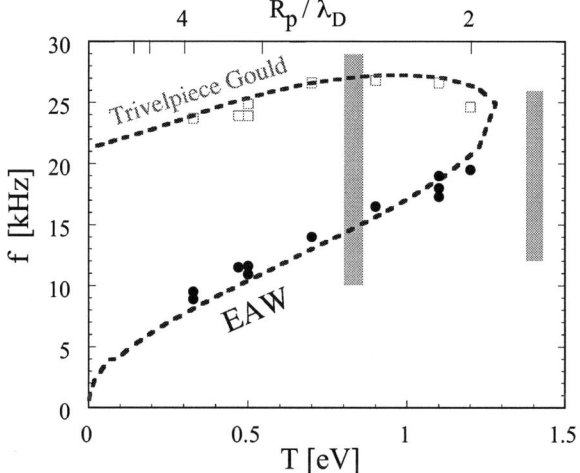

FIGURE 3. Measured plasma wave dispersion relation for fixed length with $m_z = 1$.

DISPERSION RELATION

The predicted dispersion relation of plasma waves in an infinite homogeneous plasma is shown in Fig. 2a. The upper branch is the traditional Langmuir wave starting at the plasma frequency for $k_z \lambda_D = 0$. The lower branch (EAW) first introduced by Holloway and Dorning[1] has an acoustic dispersion relation for small $k_z \lambda_D$ with a low phase velocity $v_{ph} \simeq 1.4 \bar{v}$, where $\bar{v} = (T/m)^{1/2}$. Their analysis considers a flattened particle distribution at v_{ph} eliminating the otherwise strong Landau damping.

In a trapped non-neutral plasma with a finite radial size, the perpendicular wave number is set by the radial plasma size. The fast plasma wave (TG) also has an acoustic dispersion relation and has $\omega < \omega_p$ since the radial wall short-circuits part of the electric field of the wave. In contrast, the EAW branch is essentially the same in a trapped

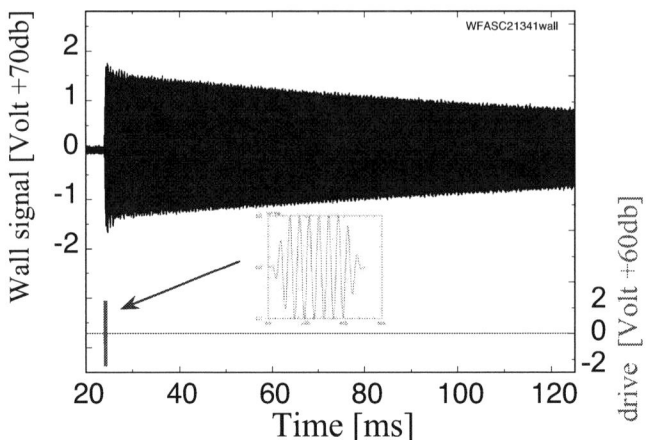

FIGURE 4. Trivelpiece-Gould wave driving burst and received wall signal.

plasma as in an unbounded plasma, as shown in Fig. 2b. For the data presented in this paper, the plasma length determines $k_z \approx m_z \pi / L_p$, where m_z is the axial mode number; $m_z = 1$ consists of half a wavelength in the trapped plasma. The plasma radius determines $k_\perp = \frac{1}{r_p}(\frac{2}{\ln(r_w/r_p)})^{1/2}$. This dispersion relation for $m_z = 1$ and variable plasma temperature results in the dashed line in Fig. 3.

When the amplitude is turned down sufficiently ($A_{\text{exc}} \sim 50$ mV), the waves are observed at only one frequency, plotted in Fig. 3 as dots (EAW) and squares (TG wave) for different temperatures. At small amplitude, these measurements are well described by the near-linear theory of Refs. [1, 2]. At temperatures above 1.3 eV, no waves are observed at comparably low excitation amplitude.

However, at larger amplitude the waves are excited over a range of frequencies and furthermore they ring at frequencies different than f_{EAW} or f_{exc} because the excitation has significantly modified the particle distribution function. The gray bar at $T = 0.8$ eV shows the range of frequencies over which a 100 cycle burst with $A_{\text{exc}} = 300$ mV resulted in a wave $f_w = f_{\text{exc}}$ ringing for hundreds of cycles. This means that at $T = 0.8$ eV a wave can be excited at "any frequency" within the vertical extent of the gray bar. Similarly, plasma waves at $T = 1.4$ eV are excited with $A_{\text{exc}} = 200$ mV for 100 cycles, past the "end of the thumb" as shown by the gray bar where no near-linear solution exists.

The drive modifies the particle distribution until the distribution becomes resonant with the drive. Wave names in these continuous regimes are ill-defined, since wave names are given for well characterized distributions such as Maxwellian or near-Maxwellian and beams.

Figure 4 shows a driving burst, consisting of 10 cycles at 21.5 kHz, applied to the excitation electrode, and the TG wave detected on a separate electrode. The received wall signal grows smoothly during the burst. Figure 5 shows a driving burst consisting of 100 cycles at 10.7 kHz with ramped amplitude to avoid exciting a TG wave. The

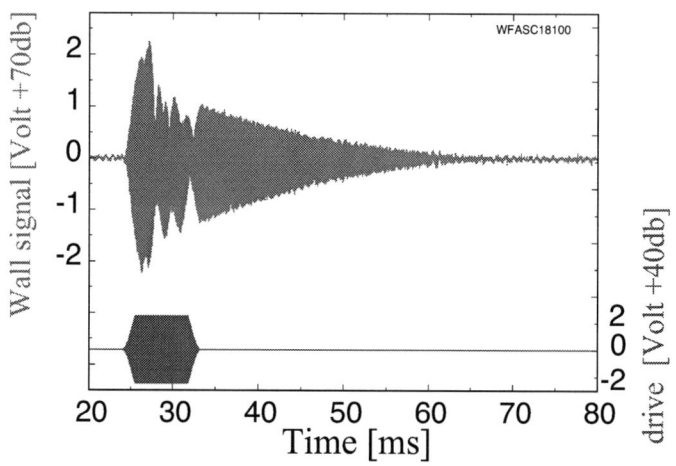

FIGURE 5. EAW driving burst and received wall signal.

EAW signal received on the wall A_w is "erratic" during the exciting burst reflecting the complicated process by which the particle distribution forms a plateau. If the driving burst is terminated after about 20 cycles, when A_w is the largest, the plasma oscillations damp within one or two wave cycles, presumably because the plateau was not formed yet. The fully-developed EAWs are observed to damp exponentially with $30 < \gamma < 3000 s^{-1}$; this is about $10\times$ faster than TG waves of comparable amplitude.

CHIRP DRIVE, EXTREME MODIFICATION OF $F(v)$

Similar plasma modes can also be excited to very large amplitude by a down-chirped frequency drive [6]. Here the chirped frequency creates extreme modification of $F(v)$, and can be tailored to support a mode at almost any frequency. We measure the parallel particle velocity distribution with a laser beam aligned along the magnetic field. Figure 6 shows an example of extreme modification of $F(v)$ from an amplitude-rounded ($A_{\text{exc}} \approx 800$ mV) burst of 14.5 cycles total chirped from $20 \rightarrow 9$ kHz, corresponding to phase velocity v_{ph1} and v_{ph2} in Fig. 6. The glitch just to the left of $v = 0$ and the 10% left-right sensitivity differences are laser-cooling artifacts. Here, the original Maxwellian distribution has been essentially split into two counter-propagating distributions, each supporting an EAW-like wave on the "inside" of $F(v)$. This wave rings at $f_w = 9.9$ kHz, with an amplitude giving $\delta n/n \sim 0.3$, with very weak damping $\gamma \approx 0.5$ sec^{-1}, $\gamma/\omega \sim 10^{-5}$. Similar frequency sweeping of phase space structures has been studied theoretically and numerically [8, 9], and driven phase space "holes" have been numerically explored [9, 10] as an explanation for observations of almost-undamped plasma waves in trapped systems [11].

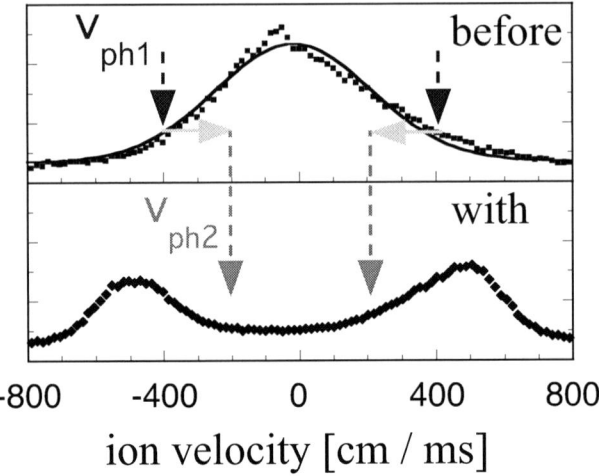

FIGURE 6. Particle distribution before wave and with plasma wave excited by a chirp sweeping from v_{ph1} down to v_{ph2}.

DISCUSSION

We have observed new, near-linear plasma waves with a phase velocity slow enough to be located in the bulk of the particle velocity distribution. At small amplitude, the experimentally observed standing wave in a magnesium ion plasma confirms the EAW theory concept of Holloway and Dorning, that is a local flattening of the particle distribution around the phase velocity suppressing Landau damping. At moderate and large amplitude, the wave can be excited at the small amplitude theory frequency, but also over a wide continuum range of frequencies. Here the wave driver modifies the particle velocity distribution until the distribution becomes resonant with the driver.

ACKNOWLEDGMENTS

This work was supported by National Science Foundation Grant No. PHY0354979 and NSF/DOE grant PHY0613740.

REFERENCES

1. J.P. Holloway and J.J. Dorning, Phys. Rev. A **44**, 3856 (1991).
2. F. Valentini, T.M. O'Neil and D.H.E. Dubin, Phys. Plas. **13**, 052303 (2006).
3. A.A. Kabantsev, F. Valentini, and C.F. Driscoll, in NNP VI (edited by M. Drewsen et al., eds.), AIP Conf. Proc. **862**, 13 (2006); F. Valentini, T.M. O'Neil and D.H.E. Dubin, Phys. Plas. **14**, 052103 (2007).

4. D.S. Montgomery *et al.*, Phys. Rev. Lett. **87**, 155001 (2001).
5. N.J. Sircombe, T.D. Arber and R.O. Dendy, Plas. Phys. and Control. Fusion **48**, 1141 (2006).
6. W. Bertsche, J. Fajans and L. Friedland, Phys. Rev. Lett. **91**, 265003 (2003); F. Peinetti *et al.*, Phys. Plasmas **12**, 062112 (2005).
7. B. Afeyan *et al.*, Proc. Inertial Fusion Sciences and Applications 2003 (B. Hamel, *et al.*, eds.), Monterey: American Nuclear Society (2004), p. 213.
8. H. Schamel, Phys. Plasmas **7**, 4831 (2000).
9. D.Y. Ermin and H.L. Berk, Phys. Plasmas **9**, 772 (2002); H.L. Berk, B.N. Breizman and N.V. Petviashvilli, Phys. Lett. A **234**, 213 (1997).
10. F. Peinetti, F. Peano, G. Coppa, and J. Wurtele, J. Comp. Physics **218**, 102 (2006).
11. L. Friedland *et al.*, Phys. Plasmas **11**, 4305 (2004).

Excitation of high order diocotron modes in the ELTRAP device

G. Bettega*, B. Paroli†, R. Pozzoli* and M. Romé*

*I.N.F.N. Sezione di Milano and Dipartimento di Fisica, Università degli Studi di Milano, Via Celoria 16, Milano, I-20133, Italy
†Dipartimento di Fisica, Università degli Studi di Milano, Via Celoria 16, Milano, I-20133, Italy

Abstract. The excitation of $l = 2, 3$ diocotron modes in a trapped pure electron plasma with external time varying electrostatic wall asymmetries is experimentally investigated. Direct measurements of the resistive destabilization of the $l = 2$ diocotron mode are presented. The experimental results are well described by a two dimensional (2D) cold fluid model.

INTRODUCTION

Due to their exceptional confinement properties, cylindrical Penning-Malmberg traps for the storage of non-neutral plasmas are well suited devices for the investigation of basic Plasma Physics [1]. In this context an accurate knowledge of the elementary processes of collective modes excitation and damping has a great importance both manipulation and for diagnostic purposes. The trap ELTRAP [2] is designed for the long-time, steady state confinement of a low density ($n \approx 10^{12} - 10^{13}\,\mathrm{m}^{-3}$) and temperature ($T \approx 1 - 10\,\mathrm{eV}$) pure electron plasma (see Fig. 1 for a schematic configuration of the apparatus). An electron column with radius $R_P \approx 1 - 3\,\mathrm{cm}$, length $L_P \approx 50\,\mathrm{cm}$, is axially confined under high vacuum conditions ($p \approx 10^{-9} - 10^{-8}\,\mathrm{mbar}$) within a stack of hollow conducting cylindrical electrodes by two fixed negative voltages ($|V_C| = 50 - 100\,\mathrm{V}$), and a uniform axial magnetic field (up to $B = 0.2\,\mathrm{T}$) which induces a bulk equilibrium rotation of the plasma, providing radial confinement. A uniform density electron plasma column rotates at the diocoton frequency, defined as $\omega_D = en/\varepsilon_0 B$. The electron plasma is generated by a W spiral cathode which continuously emits particles. The device operates following repeated cycles of plasma injection, hold (trapping) and dump: machine cycle is described in [2]. The plasma evolution is monitored through electrostatic diagnostic, measuring the induced currents at the walls when particles are trapped, and through optical diagnostic at the end of the machine cycle, when plasma is dumped onto a positively biased ($\approx 10 - 15\,\mathrm{kV}$) phosphor screen. At this time a short (1ms) light pulse is produced and colleced by a CCD camera, obtaining a 2D longitudinally averaged picture of the charge density distribution. A 2D cold fluid model [3] based on the drift-Poisson equations is usually employed to describe an important class of waves, the diocotron modes, running on the plasma column as incompressible density (and potential) waves, with pure azimuthal spatial dependence of the type $\delta n^{(l)}(r,t)\exp(il\theta)$ (l is the azimuthal wavenumber). In previous experiments the $l = 2$ diocotron mode was excited using static wall asymmetries applied for a fraction of the equilibrium rotation period $2\pi/\omega_D$ and

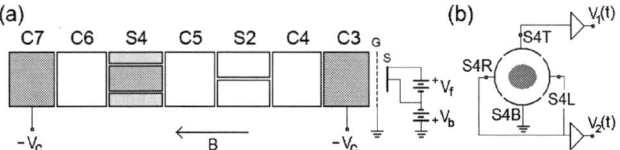

FIGURE 1. (a) Schematic of the ELTRAP device. The 7 electrodes used in the experiments are shown. S2 and S4 cylinders are azimuthally sectored. Source assembly is shown, made of a hot filament S, a power source V_f, a bias power supply V_b and an extraction grid G (b): electrical connections for the measurement of $l = 1$ mode (single dipole antenna) and $l = 2$ diocotron mode (two electrically connected patches).

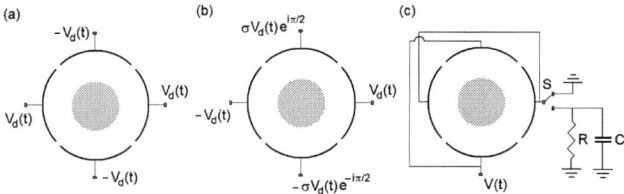

FIGURE 2. (a) Quadrupole perturbation obtained with the voltage signal $V_d(t) = V_d \sin(\omega_d t)$ (b) Rotating wall perturbation: $V_d(t)$ is applied to the 4 patches of the S4 electrode, phase shifted by $\pi/2$. $\sigma = \pm 1$ determines the sense of rotation of the overall field asymmetry (c) Setup for the investigation of the resistive wall effect on the the $l = 2$ diocotron mode. R value is adjustable and S is a computer controlled switch. Two patches are used as receiving antennas.

then removed, with the aim to measure mode damping non-linear oscillations [4]. Also long time ($\gg 2\pi/\omega_D$) and higher order (tripolar) static asymmetries were used, in order to observe static deformed equilibria [5]. In both cases wall asymmetries matching the desired deformations were used. In the following it will be shown that it is possible to excite $l = 2,3$ diocotron modes using time varying, resonant, wall perturbations, and, in the case of the $l = 3$ mode using the rotating wall (RW) technique. Consider a step like equilibrium density profile $n_0(r) = n_0[1 - H(r - R_P)]$: diocotron waves are localized at the plasma surface $r = R_P$ and propagate with frequencies $\omega^{(l)} = \omega_D[l - 1 + (R_P/R_W)^{2l}]$. Using the linearized drift-Poisson equations the perturbation of the potential of the lth mode reads

$$\frac{1}{r}\frac{\partial}{\partial r} r \frac{\partial \delta\phi^{(l)}(r,s)}{\partial r} - \frac{l^2}{r^2}\delta\phi^{(l)}(r,s) = -\frac{2il\omega_D}{s + il\omega_D}\frac{\delta\phi^{(l)}(r,s)}{r}\delta(r - r_P), \quad (1)$$

in which the perturbations have been expanded in azimuthal Fourier modes, and time behavior is studied in the Laplace domain.

RESONANT EXCITATION OF THE $l = 2,3$ DIOCOTRON MODES

At first the configuration of Fig. 2-(a) has been used in order to excite the $l = 2$ diocotron mode [6]: in this case the field perturbation inside the trap contains only $l = 4k + 2$

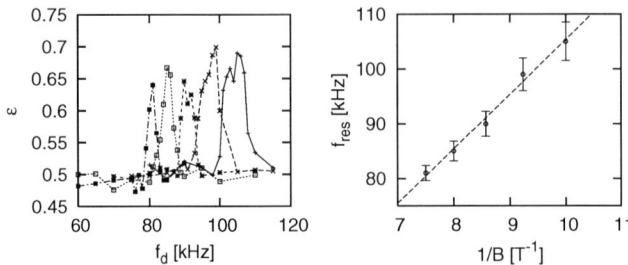

FIGURE 3. **Left:** deformation of the plasma cross section ($l = 2$ mode amplitude) as a function of the drive frequency $\omega_d/2\pi$, for different values of the axial confining magnetic field, and for a fixed perturbation time ($90\,\mu s$). The drive amplitude i $V_d = 0.6\,\text{V}$, much less than the characteristic radial potential of the plasma, estimated with $\phi \approx e n_0/\varepsilon_0 R_P^2$. **Right:** resonance frequency versus confining magnetic field. The straight line is a B^{-1} fit.

modes, being k an integer. The electrostatic equation is solved in the plasma ($0 \le r < R_P$) and vacuum ($R_P < r \le R_W$) regions, the solutions are matched at $r = R_P$, then after application of the specific boundary condition, the following result is obtained, in the case of a continuously applied drive:

$$\delta n^{(2)}(r,\theta,s) = -\frac{4n_0 V_d}{\pi B}\left(\frac{r}{R_W}\right)^2 \frac{\delta(r-R_P)}{r}\exp(i2\theta)\frac{\omega_d}{(s^2+\omega_d^2)(s+i\omega^{(2)})}. \quad (2)$$

This equation shows that at resonance, i.e. when $\omega_d = \omega^{(2)}$, the $l = 2$ mode amplitude grows linearly in time, because of the presence of a couple of pure imaginary poles with double multiplicity. In the experiments the plasma develops an evident elliptical deformation when the drive frequency ω_d is close to the estimated $l = 2$ mode frequency. Such an effect has been revealed using the CCD camera, measuring the eccentricity of the plasma edge, experimentally taken as $l = 2$ mode amplitude. The drive frequency has been also varied searching for resonances, using a fixed time duration of the wall asymmetry ($90\,\mu s$). The results are shown in Fig. 3-left: broadened resonance curves have been drawn, with a maximum plasma deformation localized around $\omega_d \approx \omega^{(2)}$. The resonant frequency scales as B^{-1}, as expected in the case of a near 2D $\mathbf{E} \times \mathbf{B}$ dynamics: see Fig. 3-right. A first evidence of non linear effects not included in the model comes from the time behavior of the mode amplitude versus the time duration of the drive, measured for different drive frequencies. Indeed, according to the calculated plasma response the mode amplitude has to grow linearly near resonance, however after an initial linear growth, abrupt jumps have been experimentally found, as shown in Fig. 4-left. They are due to the auto-resonant coupling between the drive and the $l = 2$ mode: in practice the drive is fixed at ω_d, while $\omega^{(2)}$ spontaneously shifts (decreases) because of the non-linear dependence on the (increasing) deformation: when $\omega_d = \omega^{(2)}$ phase locking occurs and an abrupt change in the mode amplitude occurs. Jumps anticipate if the drive frequency is chosen closer to the exact value of the resonance frequency. At resonance non-linear structures have been observed due to high plasma deformation and radial particle transport [4], as shown in Fig. 4-right.

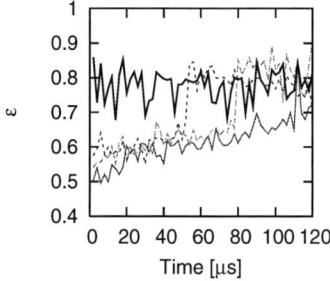
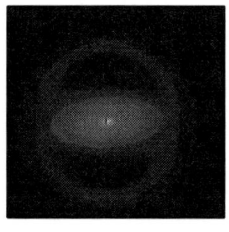

FIGURE 4. Left: deformation of the plasma versus time. The drive is continuously applied. Drive amplitude $V_d = 0.6$V. Thin solid, long-dashed, short-dashed, thick solid lines correspond to $f_d = 92, 94, 64, 100$kHz, respectively. In this experiment the resonance frequency is 100kHz. **Right:** highly deformed plasma cross section at resonance, with *cat's eye* structure.

The resonant excitation technique just described can be extended to higher order diocotron modes: for $l > 2$ an increasing number of azimuthal electrodes is required in order to build up a launching antenna matching the spatial behavior of the wave. Actually it is possible to excite the $l = 3$ mode, corresponding to a tripolar deformation of the plasma cross section, using only 4 patches and a rotating wall field [7]. Refer to the setup in Fig. 2-(b). The RW perturbation has a dominant dipole component, rotating in clockwise or counterclockwise sense, depending on the value of the σ parameter. Actually Fourier expanding the perturbation non zero odd components of the electrostatic potential are found: each of them is in principle able to excite the corresponding diocotron mode. Fourier (space) and Laplace (time) transforming the RW boundary condition and using the electrostatic equation, the response of the odd $l = 2k+1$ diocotron mode in the presence of the RW field, in terms of density perturbation, reads:

$$\delta n^{(2k+1)}(r,s) = -\frac{2\varepsilon_0 V_d \omega_D}{\pi e} \left(\frac{r}{R_W}\right)^{2k+1} \frac{\delta(r-R_P)}{r} \cdot \frac{1+i(-1)^k}{s^2+\omega_d^2} \cdot \frac{s+i\omega_d\sigma(-1)^k}{s+i\omega^{(2k+1)}}, \quad (3)$$

in which s is the complex oscillation frequency. Poles in the plasma response are physically related to the presence of the free running wave at frequency $\omega^{(2k+1)}$, and to the presence of a forced oscillation at ω_d. Depending on the σ value a couple of pure imaginary poles can be cancelled out, or doubled. In particular considering the case $k = 2$ ($l = 3$ mode) it is found that in presence a counter-rotating ($\sigma = 1$) and resonant drive ($\omega_d = \omega^{(3)}$), the amplitude of the $l = 3$ wave linearly grows in time because of the presence of a couple of pure imaginary poles with double multiplicity. As in the case of the resonant excitation of the $l = 2$ mode, broadened resonance curves have been experimentally obtained, applying the counter-rotating wall drive for 100ms at a fixed magnetic field, and varying the drive frequency: results are shown in Fig. 5-left. In case of high excitation non linear structures are formed (Fig. 5-right). If a resonant ($\omega_d = \omega^{(3)}$) and co-rotating drive ($\sigma = -1$) is applied, no mode excitation is found in the experiments, according to the presence of a couple of pure imaginary poles, which means stable oscillations at constant amplitude.

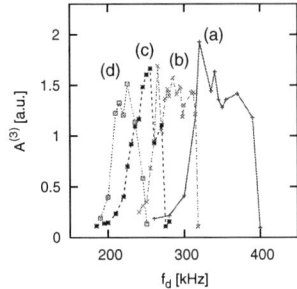 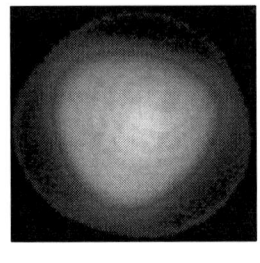

FIGURE 5. **Left:** resonance curves obtained with (a) $B = 0.067$ T, (b) $B = 0.1$ T, (c) $B = 0.117$ T, (d) $B = 0.133$ T. The drive amplitude is $V_d = 2.2$ V. The mode amplitude is represented by $A^{(3)}$, the Fourier component of the deformation of the plasma edge. **Right:** formation of non-linear structures close to the resonance condition ($B = 0.067$ T and $\omega_d/2\pi = 320$ kHz.)

RESISTIVE EXCITATION OF THE $l = 2$ DIOCOTRON MODE

Diocotron modes are destabilized when dissipation is introduced in the system. First experimental evidence of the negative energy behavior [8] came from the measurement of the resitive destabilization of the fundamental $l = 1$ diocotron mode: in the experiments described in [9] energy dissipation was introduced through an external impedance. In [10] decreasing values of intrinsic damping rate of the $l = 2$ mode were measured as a function of an increasing external resistive load: in such a way the negative energy behavior of the quadrupole perturbation was indirectly inferred, without measuring any growth. First direct measurement of the resistive wall effect on the $l = 2$ has been recently obtained [11], using the experimental setup shown in Fig. 2-(c): a resistor is attached to a couple of electrically connected azimuthal electrodes by a computer controlled swicth. C represents the trasmission line capacitance. The remaining patches of the composite S4 electrode work as receiving antenna for the mode diagnostic. After injection the plasma freely evolves and thermalizes, then the resistive load is inserted for 0.6 s, finally that walls are grounded. The $l = 2$ mode amplitude and frequency are measured through the charge signal on the receiving antennas. A typical results is shown in Fig. 6. The external resistive load determines an exponential growth of the mode amplitude, with growth rate $\gamma^{(2)}$ which depends on the resistor value and on the total line capacitance. In order to describe the experimental observations a 2D linear model has been developed, which takes into account the coupled plasma-external network dynamics through a time varying boundary condition on the resistive walls, determined by the instantaneous response of the RC circuit. Calculation is performed in space and time Fourier domains, i.e. using $s = -i\omega$ in the electrostatic equation, because only the regime response survives after the initial RC transient. A complex frequency shift $\delta\omega^{(2)}$ is obtained, whose positive imaginary part corresponds to an exponential mode growth rate:

$$\gamma^{(2)} = Im[\delta\omega^{(2)}] = \frac{8\omega_D\varepsilon_0 L_S^2}{\pi C L_P}\left(\frac{R_P}{R_W}\right)^4 \frac{\omega^{(2)}RC}{1+\omega^{(2)2}R^2C^2} > 0, \qquad (4)$$

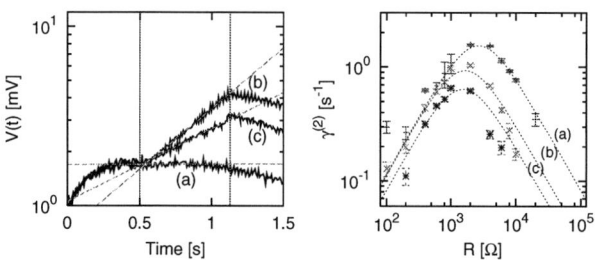

FIGURE 6. Left: induced current on the antennas transformed in to a voltage by a low noise transimpedance amplifier with gain $1.1 \times 10^6 \, \mathrm{VA^{-1}}$, transforming induced currents in an output voltage. Measured line capacitance $C = 906 \, \mathrm{pF}$. (a) $R = 0 \, \Omega$, $\gamma^{(2)} = 8.23 \times 10^{-3} \, \mathrm{s^{-1}}$ (b) $R = 2 \, k\Omega$, $\gamma^{(2)} = 1.56 \, \mathrm{s^{-1}}$ (c) $R = 8 \, k\Omega$, $\gamma^{(2)} = 9.34 \times 10^{-1} \, \mathrm{s^{-1}}$ **Right:** measured mode growth rates $\gamma^{(2)}$ as a function of the resistive load for different line capacitance. (a) $C = 906 \, \mathrm{pF}$ (b) $C = 1.37 \, \mathrm{nF}$ (c) $C = 1.84 \, \mathrm{nF}$.

in which $L_S = 15 \, \mathrm{cm}$ is the axial length of the sectored cylinder. A good agreement between the experimental measured growth rates and the theoretical prediction has been found, as shown Fig. 6-right for three different values of the transmission line capacitance.

Concluding it has been shown that within the limit of a cold fluid 2D linear approximation, a pure electron plasma trapped in a Penning-Malmberg trap behaves like a simple oscillator, with poles corresponding to the frequency of diocotron modes. Each mode reacts to an external resonant drive with a linearly increasing amplitude, as experimentally found in the case of the excitation of the $l = 2$ with a quadrupole field asymmetry, and in the case of the resonant excitation of the $l = 3$ mode, with the residual $l = 3$ harmonic component of a rotating wall perturbation. First direct measurements of the resistive destabilization of the $l = 2$ mode are presented. Also in this case the cold fluid 2D linear model based on the drift Poisson equations with time dependent boundary conditions is able to describe the main features of the experimental observations.

REFERENCES

1. J. H. Malmberg, J. S. deGrassie, Phys. Rev. Lett. **35**, 577 (1975).
2. M. Amoretti, G. Bettega, F. Cavaliere, M. Cavenago, F. De Luca, R. Pozzoli and and M. Romé, Rev. Scient. Instr. **74**, 3991 (2003).
3. R. C. Davidson, *An Introduction to the Physics of Nonneutral Plasmas* (Addison-Wesley, Redwood City, 1990).
4. A. C. Cass, "Experiments on vortex symmetrization in magnetized electron columns," Ph. D. dissertation, University of California at San Diego (1998).
5. R. Chu, J. S. Wurtele, J. Notte, A. J. Peurrung and J. Fajans, Phys. Fluids B **5**, 2378 (1993).
6. G. Bettega, F. Cavaliere, B. Paroli, M. Cavenago, R. Pozzoli and M. Romé, Phys. Plasmas **14**, 102103 (2007).
7. G. Bettega, B. Paroli, R. Pozzoli and M. Romé, submitted to Jour. Appl. Phys.
8. R. J. Briggs, J. D. Daugherty, and R. H. Levy, Phys. Fluids **13**, 421 (1970).
9. W. D. White, J. H. Malmberg, and C. F. Driscoll, Phys. Rev. Lett. **49**, 1822 (1982).
10. N. S. Pillai and R. W. Gould, Phys. Rev. Lett. **73**, 2849 (1994).
11. G. Bettega, B. Paroli, M. Cavenago, R. Pozzoli and M. Romé, Phys. Plasmas **15**, 032102 (2008).

Using variable-frequency asymmetries to probe the magnetic field dependence of radial transport in a Malmberg-Penning trap

D.L. Eggleston

Occidental College, Los Angeles, California, USA

Abstract. A new experimental technique is used to study the dependence of asymmetry-induced radial particle flux Γ on axial magnetic field B in a modified Malmberg-Penning trap. This dependence is complicated by the fact that B enters the physics in at least two places: in the asymmetry-induced first order radial drift velocity $v_r = E_\theta/B$ and in the zeroth order azimuthal drift velocity $v_\theta = E_r/B$. To separate these, we employ the hypothesis that the latter always enters the physics in the combination $\omega - l\omega_R$, where $\omega_R = v_\theta/r$ is the column rotation frequency and ω and l are the asymmetry frequency and azimuthal mode number, respectively. Points where $\omega - l\omega_R = 0$ are then selected from a Γ vs r vs ω data set, thus insuring that any function of this combination is constant. When the selected flux Γ_{sel} is plotted versus the density gradient, a roughly linear dependence is observed, showing that this selected flux is diffusive. This linear dependence is roughly independent of the bias of the center wire in our trap ϕ_{cw}. Since in our experiment ω_R is proportional to ϕ_{cw}, this latter point shows that our technique has successfully removed any dependence on ω_R and its derivatives, thus confirming our hypothesis. The slope of a least-squares fitted line through the Γ_{sel} vs density gradient data then gives the diffusion coefficient D_0 under the condition $\omega - l\omega_R = 0$. Varying the magnetic field, we find D_0 is proportional to $B^{-1.33 \pm 0.05}$, a scaling that does not match any theory we know. These findings are then used to constrain the form of the empirical flux equation. It may be possible to extend this technique to give the functional dependence of the flux on $\omega - l\omega_R$.

Keywords: non-neutral plasma, asymmetry-induced transport, magnetic field dependence
PACS: 52.27.Jt, 52.55.Dy, 52.25.Fi

INTRODUCTION

The Malmberg-Penning non-neutral plasma trap continues to be of interest both as a platform for basic plasma physics studies and for its applications in charged particle storage and manipulation. While it is well established that electric and magnetic fields that break the cylindrical symmetry of these traps produce radial transport, a full understanding of the transport remains elusive. Indeed, our work, which focuses on the transport produced by applied electric asymmetries with frequency ω and axial and azimuthal wavenumbers n and l, has revealed serious discrepancies between experiment[1] and some of the predictions of theory[2].

Faced with these discrepancies, we have turned to developing an empirical model of the transport with an eye toward providing guidance for further theoretical development. A basic issue in this program is determining the magnetic field dependence of the transport. Although this scaling has been studied before[4, 5], there is no consensus of results. This may be due to the fact that the magnetic field B enters the transport physics in at least two ways. Firstly, in the zeroth order azimuthal $E \times B$ drift produced by the radial electric field $v_\theta = E_r/B$. This causes the particle guiding centers to drift around the trap axis with angular frequency $\omega_R = v_\theta/r$. Secondly, the magnetic field enters in the first order radial $E \times B$ drift produced by the applied asymmetry $v_r = E_\theta/B$. It is this drift which is responsible for the radial transport of particles. This dual dependence on magnetic field can be seen, for example, in the expression for the flux Γ from the plateau regime of resonant particle transport theory[2].

In this paper we apply a new experimental technique to remove the ω_R dependence and thus isolate any remaining magnetic field dependence. The technique is based on the hypothesis that the asymmetry frequency ω and ω_R always enter the transport physics in the combination $\omega - l\omega_R$. We then select from a Γ vs r vs ω data set those points where $\omega - l\omega_R = 0$, thus insuring that any function of this combination is constant. When the selected flux Γ_{sel} is plotted versus the density gradient ∇n_0, a roughly linear dependence is observed, showing that this selected flux is at least partially diffusive. This linear dependence is roughly independent of the center wire bias ϕ_{cw}. Since in our experiment $\omega_R \propto \phi_{cw}$, this latter point shows that our technique has successfully removed any dependence on ω_R and its derivatives, thus confirming our hypothesis. The slope of a least-squares fitted line through the Γ_{sel} vs ∇n_0 data then gives the diffusion coefficient under the condition $\omega - l\omega_R = 0$ which we call D_0. Varying the magnetic field, we find $D_0 \propto B^{-1.33 \pm 0.05}$. We then use these findings to constrain the form of the empirical flux equation[3].

EXPERIMENTAL DEVICE

Our transport studies are performed in the modified Malmberg-Penning trap shown in shown in Fig. 1. As in the standard trap design, a uniform axial magnetic field provides radial confinement of injected electrons, while negatively biased end cylinders (the injection gate and dump gate) provide axial confinement. Our device also operates in the standard inject-hold-dump cycle. A cycle begins by grounding the injection gate which allows electrons from the gun to flow into the central region. This injection gate is then returned to a negative bias which traps the electrons. After a chosen period of time, the dump gate is grounded and the electrons leave the trap and hit a positively biased phosphor screen. Analysis of the images on this screen provides the primary diagnostic.

The principal modification in our device is replacing the usual plasma column with a biased wire running along the axis of the trap. The wire provides a radial electric field to replace the field normally produced by the plasma column and allows the injected low density electrons to have the same zeroth-order dynamical motions (axial bounce and azimuthal $E \times B$ drift motions) as in a standard trap. The lower density (10^5 cm^{-3}) and high temperature (4 eV) of the electrons give a Debye length larger than the trap radius. Under these conditions, potentials in the plasma are essentially the vacuum potentials and previously encountered[6] complications due to collective effects are minimized[1]. Our design also allows the drift rotation frequency $\omega_R(r)$ to be easily adjusted by varying the center wire bias ϕ_{cw} since $\omega_R = \frac{-\phi_{cw}}{r^2 B \ln(R/a)}$ where R and a are the radii of the wall and the center wire, respectively. Despite these changes, the unperturbed confinement time has similar magnitude and shows[7] the same $(L/B)^2$ scaling found in higher

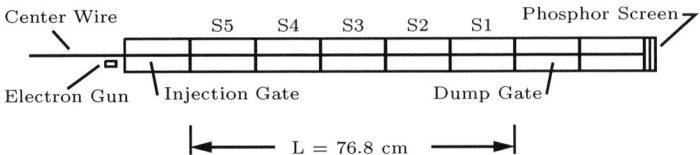

FIGURE 1. Schematic of the Occidental College Trap. The usual plasma column is replaced by a biased wire to produce the basic dynamical motions in low density electrons injected from an off-axis gun. The low density and high temperature of the injected electrons largely eliminate collective modifications of the vacuum asymmetry potential. The five cylinders (labeled S1 through S5) are divided azimuthally into eight sectors each.

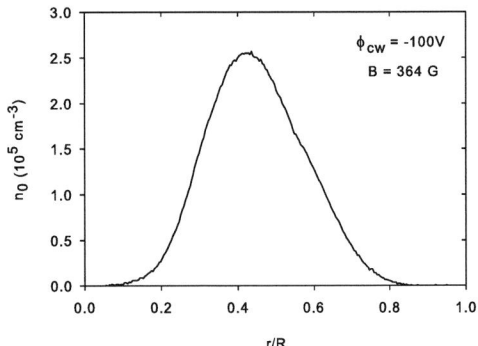

FIGURE 2. A typical density profile taken 1600 ms after injection.

density experiments, thus supporting the idea that the radial transport is primarily a single particle process and confirming the relevance of our experiments to standard trap physics.

A unique feature of our device is that the entire confinement region is sectored (five cylinders, labeled S1 through S5 in Fig. 1, with eight azimuthal divisions each). This allows us to apply a simple, known asymmetry by selecting the amplitude and phase of the voltages applied to each sector to produce a helical standing wave of the form $\phi(r,\theta,z,t) = \phi_W \left(\frac{r}{R}\right)^l \cos\left(\frac{n\pi z}{L}\right)\cos(l\theta - \omega t)$ where ϕ_W is the asymmetry potential at the wall (typically 0.2 V), R is the wall radius (3.82 cm), L is the length of the confinement region (76.8 cm), n and l are the axial and azimuthal Fourier mode numbers, respectively, and z is measured from one end of the confinement region. For these experiments $n = l = 1$ and the relative phases of the applied voltages are adjusted so that the asymmetry rotates in the same direction as the zeroth-order azimuthal $E \times B$ drift.

Data acquisition for these transport studies can be summarized as follows; details have been given elsewhere [1, 8]. Electrons injected into the trap from an off-axis gun are quickly dispersed into an annular distribution. At a chosen time (here, 1600 ms after injection), the asymmetries are switched on for a period of time δt (here, 100 ms) and then switched off. At the end of the experiment cycle, the electrons are dumped axially onto a phosphor screen and the resulting image is digitized using a 512 × 512 pixel charge-coupled device camera. A radial cut through this image gives the density profile $n_0(r)$ of the electrons. A typical profile is shown in Fig. 2. Shot-to-shot variation in the number of injected electrons is less than 1% and the data is very reproducible. Calibration is provided by a measurement of the total charge being dumped. Profiles are taken both with the asymmetry on and off, and the resulting change in density $\delta n_0(r)$ is obtained. The background transport is typically small compared to the induced transport and is subtracted off. If the asymmetry amplitude is small enough and the asymmetry pulse length δt short enough, then $\delta n_0(r)$ will increase linearly in time [8]. We may then approximate $dn_0/dt \simeq \delta n_0(r)/\delta t$ and calculate the radial particle flux $\Gamma(r)$ (assuming $\Gamma(r=a) = 0$):

$$\Gamma(r) = -\frac{1}{r}\int_a^r r'dr' \cdot \frac{dn_0}{dt}(r') \tag{1}$$

Here a is the radius of the central wire (0.178 mm). The entire experiment is then repeated for a series of asymmetry frequencies ω and the resulting flux versus radius and frequency data saved for analysis.

EXPERIMENTAL RESULTS

It is easy to show experimentally that the transport depends separately on both ω and ω_R and that the form of the transport equation is more complicated than a simple Fick's Law dependence $\Gamma = -D\nabla n_0$. Some typical data is shown in Fig. 3. In Fig. 3a we plot the radial particle flux Γ versus radius r for three representative asymmetry frequencies to illustrate the dependence on ω. In Fig. 3b the same data is plotted versus density gradient ∇n_0 to show that there is no simple relationship between Γ and ∇n_0. Similar plots holding ω constant and varying ϕ_{cw} (and thus ω_R) demonstrate the dependence of the flux on ω_R[3].

We now apply the hypothesis that ω and ω_R always enter the transport physics in the combination $\omega - l\omega_R$. We take Γ vs r data for a number (typically 26) of asymmetry frequencies ω. Since $l = 1$ in our experiments, these frequencies are chosen to be within the range of ω_R values, i.e., $\omega_R(R) < \omega < \omega_R(a)$. We then select from this Γ vs r vs ω dataset those points where $\omega - l\omega_R = 0$, thus insuring that any function of this combination is constant. We do this as follows: for each experimental value of ω, we determine the radial position r_{sel} where $\omega - l\omega_R = 0$, interpolating between data points if necessary. We then take from the Γ vs r data for that ω the single flux value Γ_{sel} that occurs at r_{sel}. After this is repeated for each ω, we have Γ_{sel} vs r_{sel} with r_{sel} spanning the range of radius values. Since the plasma parameters are independent of ω, ∇n_0 does not change with ω and we can also form Γ_{sel} vs ∇n_0.

When the selected flux is plotted versus the density gradient ∇n_0, a roughly linear dependence is observed and this dependence is roughly independent of the center wire bias ϕ_{cw}. Typical data is shown in Fig. 4a. The linearity of the plot shows that the selected flux has the form $\Gamma_{sel} = m\nabla n_0 + \Gamma_0$, where m and Γ_0 are constants. In particular, m and Γ_0 are not functions of ω or ω_R. The first follows from the fact that the data points in Fig. 4a are all at different frequencies and the second follows from the lack of dependence on

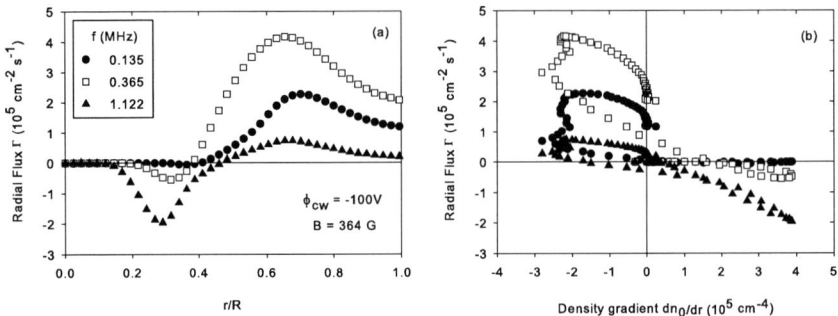

FIGURE 3. (a) Plot of typical flux versus scaled radius data for three representative asymmetry frequencies. (b) Plot of the same flux data versus density gradient. The number of plotted points has been adjusted for clarity. The plots show that the flux depends on the asymmetry frequency and does not follow a simple Fick's Law dependence on density gradient.

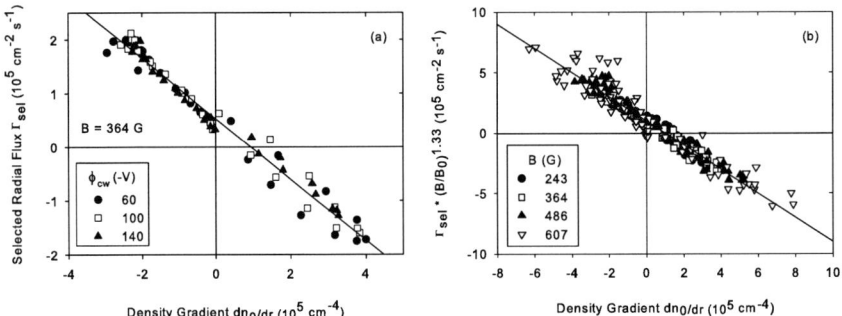

FIGURE 4. a) Selected flux versus density gradient with center wire bias as a parameter. The slope of a fitted line gives the diffusion coefficient. b) A universal curve results when the selected flux data from three center wire biases for each of four magnetic fields is multiplied by a scaling factor $(B/B_0)^{1.33}$ and plotted versus the density gradient dn_0/dr.

ϕ_{cw}. Since we know that, in general, the flux depends separately on both ω and ω_R, the independence of Γ_{sel} on these quantities supports our hypothesis that they enter the physics only in the combination $\omega - l\omega_R$. We also note that, since the points in Fig. 4a come from different radii, m and Γ_0 are not strong functions of r either, although the deviations from linearity may indicate a weak dependence on r.

Finally, the slope of a least-squares fitted line to the plot in Fig. 4a then gives the quantity m. For four values of magnetic field spanning 243-607 G, we find[3] $m \propto B^{-1.33 \pm 0.05}$. A similar procedure using the y-intercept of the fitted lines gives $\Gamma_0 \propto B^{-1.13 \pm 0.10}$.

DISCUSSION

The magnetic field scalings for m and Γ_0 are similar enough to consider a common scaling for both. This is of interest for comparison with the common theoretical form for the flux. In Fig. 4b we apply a scaling of $B^{1.33}$ to all of our data (three center wire biases for each of four magnetic fields) and obtain a universal curve of the form $(B/B_0)^{1.33}\Gamma_{sel} = -D_0(\nabla n_0 + f_0)$, where $B_0 = 233$ G is a conveniently selected constant. A least-squares fit to the scaled data gives $D_0 = 1.00$ cm^2 s^{-1} and $f_0 = -1.01 \times 10^5$ cm^{-4}. This magnetic field scaling does not match the theoretical B^{-2} plateau regime scaling or the more complicated B-scaling of the banana regime[2], or any other theoretical scaling of which we are aware.

Of course, our universal curve only gives the flux for points where $\omega - l\omega_R = 0$. It does, however, allow us to say something about the form of the general flux equation. Our data tell us that the general flux must be a function of $\omega - l\omega_R$ and that the flux equation must reduce to the equation for Γ_{sel} when $\omega - l\omega_R = 0$. Without further information, we must thus allow both D_0 and f_0 to become functions of $\omega - l\omega_R$:

$$\Gamma = -(B_0/B)^{1.33} D(\omega - l\omega_R)[\nabla n_0 + f(\omega - l\omega_R)] \qquad (2)$$

where $D(\omega - l\omega_R = 0) \equiv D_0$ and $f(\omega - l\omega_R = 0) \equiv f_0$.

CONCLUSION

We have applied a new experimental technique to study the magnetic field dependence of asymmetry-induced transport in a modified Malmberg-Penning trap. The technique allows us to remove the ω_R-dependence from our data and thus isolate the remaining magnetic field dependence. The technique works reasonably well and gives a diffusion coefficient that scales like $B^{1.33}$. This scaling does not match that of any known theory.

ACKNOWLEDGMENTS

This material is based upon work supported by the Department of Energy under award number DE-FG02-06ER54882. The author acknowledges contributions by J. M. Williams.

REFERENCES

1. D.L. Eggleston and B. Carrillo, Phys. Plasmas **10**, 1308 (2003).
2. D. L. Eggleston and T. M. O'Neil, Phys. Plasmas **6**, 2699 (1999).
3. D.L. Eggleston and J.M. Williams, Phys. Plasmas **15**, 032305 (2008).
4. J. Notte and J. Fajans, Phys. Plasmas **1**, 1123 (1994).
5. J. M. Kriesel and C. F. Driscoll, Phys. Rev. Lett. **85**, 2510 (2000).
6. D. L. Eggleston, T. M. O'Neil, and J. H. Malmberg, Phys. Rev. Lett. **53**, 982 (1984).
7. D.L. Eggleston, Phys. Plasmas **4**, 1196 (1997).
8. D. L. Eggleston and B. Carrillo, Phys. Plasmas **9**, 786 (2002).

Turbulent Cascade in Vortex Dynamics of Magnetized Pure Electron Plasmas

Yosuke Kawai and Yasuhito Kiwamoto

Graduate School of Human and Environmental Studies, Kyoto University, Kyoto 606-8501

Abstract. Elementary processes of two-dimensional (2D) turbulence has been examined by extensive analyses of fine-scale structures in the density distribution of a magnetized pure electron plasma that evolves from an unstable initial state to a single-peaked stable distribution through successive mergers of vortex patches. The observed vortex dynamics in the real space corresponds to the spectral dynamics analyzed in the wave-number (k) space. Along with the merging process among patches, the energy spectrum $E(k)$ broadens upward and its shape shows a power-law scaling $k^{-\alpha}$ with $\alpha > 3$ in a wide wave-number region above k_{inj} corresponding to the size of the first generated patches. From the energy and enstrophy tranfer rate evaluated from the time-resolved spectra, it has been observed that the enstrophy cascades at a constant transfer rate in the region $k > 3k_{inj}$, while the energy is transferred downward at $k < k_{inj}$. This spectral dynamics is qualitatively consistent with the theoretical picture of 2D turbulence with discrepancies that the power index of $E(k)$ is larger than the theoretical prediction of $\alpha = 3$ and the enstrophy transfer rate is almost zero around k_{inj} reflecting the effect of coherent vortices.

Keywords: two-dimensional turbulence, spectral analysis
PACS: 52.27.Jt, 47.27.Jv, 02.70.Hm

INTRODUCTION

In the guiding center approximation, the macroscopic dynamics of pure electron plasmas transverse to a strong magnetic field B_0 is equivalent to the two-dimensional (2D) vortex dynamics in inviscid and incompressible fluids, through the relation $\zeta = en/\varepsilon_0 B_0$ between the electron density $n(x,y)$ and the vorticity $\zeta(x,y)$ [1]. Here, $-e$ and ε_0 are the electron charge and the dielectric constant in vacuum, respectively. Taking advantage of this equivalence, many aspects of vortex dynamics constituting fundamental processes of 2D turbulence, such as the advection, merger, filamentation of vortices, has been studied extensively by employing magnetized pure electron plasmas [1–3].

Free relaxation of 2D turbulence starting from an unstable initial distribution has also been studied, focusing on the formation of quasi-stationary states with ordered structures [1, 4, 5]. In the previous study, we have investigated the relaxation process which proceeds via stochastic mergers of coherent vortices, by applying time-resolved spectral analyses, in terms of the density transport [6]. In this paper, we extend these examinations further to explore fundamental properties of 2D turbulence in the vorticity distribution in terms of the transport of the energy and enstrophy in the wave-number (k) space [7].

Theoretical picture of 2D turbulence has been proposed by Kraichnan and Batchelor [8]. In these studies, they proposed that in an isotropic and homogeneous 2D turbulence, the enstrophy (vorticity) injected at the length scale of l_{inj} ($\propto 1/k_{inj}$) cascades at a constant transfer rate of η down to a scale of dissipation l_d ($\propto 1/k_d$) and dissipates at smaller scales by viscosity. This cascade picture of 2D turbulence leads to an energy spectrum characterized by the power-law scaling $E(k) \propto k^{-3}$ in the inertial range $k_{inj} \leq k \leq k_d$ of the wave-number space.

The cascade process of the enstrophy has been investigated experimentally by using thin layers of electrolytes [10] or soap films [11], and k^{-3} scaling has been observed. However, in these experiments, it is difficult to resolve the fine vortex structures extending to the dissipative scale, and therefore there remains some uncertainties in comparing the vortex dynamics observed in the real space with the theoretical picture of 2D turbulence described in the wave-number space. In contrast to these studies, in pure electron plasma experiments, the vorticity distribution $\zeta(x,y)$ can be observed directly in terms of the electron density distribution $n(x,y)$ down to the dissipative scale at a high signal-to-noise ratio. With this advantage, in the present work, we observe and analyze the cascade process of the free-decaying 2D turbulence over a wide range of length scales, and compare the experimental results with the Kraichnan-Batchelor theory.

EXPERIMENTAL METHOD

The experiment was carried out by using a pure electron plasma confined in a Penning-Malmberg trap with a uniform magnetic field ($B_0 = 0.048$ T) and a square-well potential. The relaxation process starts with a spontaneous formation of vortices via a nonlinear stage of the diocotron instability of a ring-shaped initial distribution [2, 6]. After a generation of an initial profile, the stochastic dynamics proceeds spontaneously while the electrons are held in the trap. The time evolution of the 2D density distribution was observed destructively by damping the electrons onto the conducting phosphor screen and digitizing the resultant luminosity distribution with a charge-coupled-device camera (CCD) providing the spatial resolution of 0.1 mm/pix at each time step of the relaxation. This destructive diagnostic requires a high reproducibility of the initial profiles, because the free relaxation of the turbulence triggered by the instability is stochastic in nature. Therefore in this experiment, in addition to technically minimizing shot-by-shot variations in the initial profiles, an ensemble-average is applied over typically 5 shots of data for each time of the observation in evaluating physical quantities characterizing the turbulence. The details of the experimental configuration and diagnostic are reported in Refs. [3, 5–7].

EXPERIMENTAL RESULTS

Vortex dynamics in 2D turbulence

The time evolution of the observed density distribution is shown in Fig. 1. Each image is denoised from instrumental noise, which is accumulated charge on CCD pixels due to a dark current, by using the wavelet-based noise extraction method [9]. The ring-shaped profile produced at 5 µs is distorted by the diocotron instability [2] within a

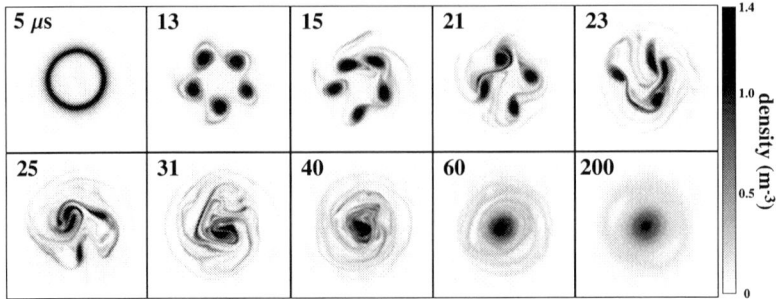

FIGURE 1. Images of the time evolution of the density distribution. The darkness is proportional to the density. The time of observation (in μs) is indicated at the upper left corner.

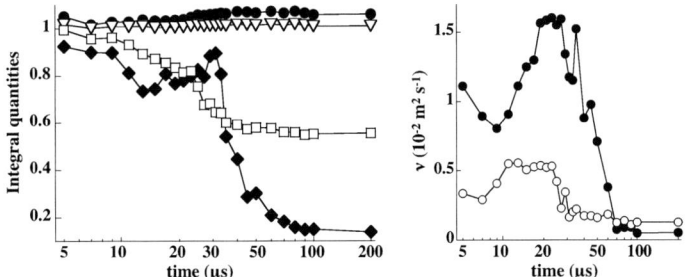

FIGURE 2. Left: Time evolution of energy E (●), total electron number N (▽), enstrophy Z_2 (□) and palinstrophy P (◆). Right: Time evolution of the viscosity coefficients evaluated from the experimental data ν (●) and calculated from the theoretical formula ν_{th} (○).

few μs, and eventually 5 high density vortex patches are generated at 13 μs. After the formation of the first vortex patches, the successive mergers among the patches proceed spontaneously up to the formation of a single vortex. This stochastic dynamics is accompanied by the filamentation of vortex structures that evolve toward smaller length scales. The concentrated patches rotate expelling filamentary structures from the central region, and finally form a single-peaked distribution surrounded by a low density halo at 200 μs.

This relaxation process is characterized by the integral quantities of the vorticity ($\propto n(x,y)$). Figure 2 (left) shows the time evolution of some integrals calculated from the measured density distribution n. The calculated integrals are the electrostatic energy (fluid kinetic energy) $E = 1/2 \int d^2\mathbf{r} n(-e\phi)$, the total electron number (total circulation) $N = \int d^2\mathbf{r} n$, the enstrophy $Z_2 = 1/2 \int d^2\mathbf{r} n^2$ and the palinstrophy $P = 1/2 \int d^2\mathbf{r} |\nabla n|^2$. Here, ϕ is the self electrostatic field calculated from $n(x,y)$. The palinstrophy is a measure of the fine-scale structures in the turbulence [8, 10].

Throughout the whole process, E and N do not show any systematic change except 5 % variations probably attributable to shot-by-shot fluctuations in the initial distributions. Therefore these integrals may be considered to be invariant. In contrast, Z_2 and P show unambiguous systematic changes. The enstrophy Z_2 undergoes a substantial decaying through the merging process, and finally goes down to 40 % of the initial value. The palinstrophy P shows a rapid increase through the merging process and is maximized at 31 μs when the filamentary structures are most conspicuous outside the high density central region as shown in Fig. 1. After the maximization, P drops steeply concurrently with the reduction of the decreasing rate of the enstrophy, which suggests an existence of a dissipative process in fine length scales.

In order to estimate the degree of enstrophy dissipation in this experiment, we evaluate an effective viscosity coefficient using the relation $DZ_2/Dt = -2\nu P$ which is derived from the Navier-Stokes equation [8, 10]. The evaluated viscosity ν is plotted as a function of time in Fig. 2 (right). ν is maximized around $t \approx 20$ μs when the density configuration changes drastically from separated vortex patches to a single-peaked distribution. In this figure, the theoretically predicted coefficient ν_{th} is also plotted, which is estimated by introducing the parameters of the present experiment into the proposed formula [12]. Though the present study is not under quasi-stationary states as assumed in the theory, the experimental evaluation ν agrees with ν_{th} within a factor of 3. This agreement may suggest that the stochastic motions of individual particles under fluctuating fields play a important role in the dissipation process of the vortex dynamics.

Spectral dynamics in 2D turbulence

To compare the experimental results to the theoretical picture of 2D turbulence, we calculate the energy spectrum in the wave-number k space from the measured density distribution. The energy spectrum $E(k)$ is determined from the Fourier transformed density distribution $n(\mathbf{k}) = \int d^2\mathbf{r} e^{-i\mathbf{k}\cdot\mathbf{r}} n(\mathbf{r})$ as

$$E(k) = \frac{1}{2}\left(\frac{e}{\varepsilon_0 B_0}\right)^2 \int_0^{2\pi} k d\varphi \frac{|n(\mathbf{k})|^2}{k^2}, \qquad (1)$$

where φ is the azimuthal angle of $\mathbf{k}$ (recall the relation $\zeta = en/\varepsilon_0 B_0$). The time evolution of $E(k)$ thus obtained is shown in Fig. 3.

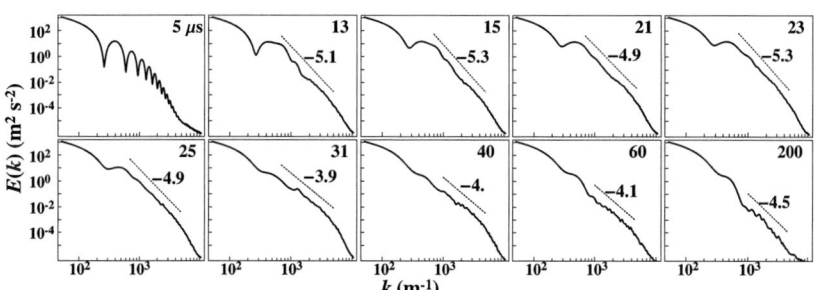

FIGURE 3. Time evolution of the energy spectrum calculated from the measured $n(\mathbf{r})$. Numbers at the upper left corner stand for the time of the observation.

When the first vortex patches are generated from the ring distribution at 13 μs, the spectrum has a local maximum around the injection scale $k = k_{inj} \approx 500$ consistent with the size of the patches. Along with the subsequent mergers between patches ($t = 13 \sim 31$ μs), the energy spectrum broadens upward and shows a power-law scaling of $\propto k^{-\alpha}$ in the wave-number region larger than k_{inj}. The slope of the spectrum in the interval $700 \leq k \leq 5000$ is drawn in Fig. 3. Throughout the merging process, the power index α remains around 5, apparently larger than the theoretically predicted value of 3 [8]. In the decaying phase associated with the reduction in the palinstrophy ($t > 31$ μs, see Fig. 2), the energy decreases steeply in the fine length scales $k > 1000$, and concentrates at $k = k_{core} \approx 300$ which corresponds to the core size in the final state ($t = 200$ μs).

The upward transfer rate of the energy $\varepsilon(k)$ through k is evaluated from the time-resolved energy spectra in Fig. 3 as

$$\varepsilon(k) = -\int_{k_{min}}^{k} dk \, \frac{\partial E(k)}{\partial t}, \tag{2}$$

where the lower integration limit k_{min} corresponds to the wall diameter. The enstrophy transfer rate $\eta(k)$ is evaluated similarly from the enstrophy spectrum $Z(k) = k^2 E(k)$. The time evolution of $\varepsilon(k)$ and $\eta(k)$ during the period when $E(k)$ shows the power-law scaling is shown in Fig. 4. Both in the energy and enstrophy, the transfer rates are maximized at 25 μs when the density configuration changes drastically.

These spectra shows the characteristic features of 2D turbulence: The enstrophy is transferred upward in the wave-number space with $k \geq k_{inj}$, while the energy cascades toward lower wave-numbers and its transfer rate is maximized around k_{core} at each time. In particular, over the wide range of $k > 1500$, $\eta(k)$ is almost constant as assumed in the 2D turbulence theory. Using the transfer rate observed at $\eta \approx (0.52 - 2.2) \times 10^{14}$ s^{-3} and the effective viscosity ν evaluated in the previous subsection, the dissipative scale l_d is estimated to be 0.57 ± 0.07 mm according to the expression $l_d \approx \eta^{-1/6} \nu^{1/2}$ [8, 10]. This length is consistent with the thickness of the filamentary structure at the end of spiral arms shown in Fig. 1, and in the wave-number space, this scale $k = k_d \approx 5500$ corresponds to the upper limit of the power-law scaling in $E(k)$ as shown in Fig. 3.

In contrast to the large wave-number region, in the intermediate scale $k_{inj} \leq k \leq 1500$ $\eta(k)$ decreases to zero around k_{inj} from a constant value at $k > 1500$, which indicates the inhibition of the enstrophy cascade. This observation corresponds to the vortex dynamics

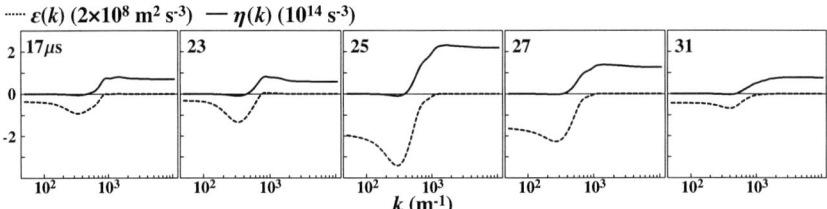

FIGURE 4. Time evolution of the upward transfer rates of the energy $\varepsilon(k)$ (dashed line) and enstrophy $\eta(k)$ (solid line) through k.

observed in Fig. 1, i.e., throughout the merging process, the vortex patches retain its coherent properties, spatial locality and high vorticity, and the filamentation of vortex structures proceeds in the outside region of the patches. The constraint of the enstrophy cascade by the coherent vortices has been observed by numerical simulations [13], and this breakdown of the cascade model is probably the reason why the power index of the observed energy spectrum $E(k)$ is larger than the theoretical prediction of $\alpha = 3$ [8].

CONCLUSION

In this paper, we have examined the relaxation process of 2D turbulence in a magnetized pure electron plasma over a wide range of length scales extending from the injection scale down to the dissipative scale, and compared the experimental results to theoretical picture of 2D turbulence. In the stage characterized by the successive mergers among vortex patches starting from the unstable initial density profile, the observed density distribution exhibits turbulent characteristics. While the energy is transfered downward, the enstrophy undergoes an upward transport starting from the injection wave-number k_{inj}. In smaller length scales with $k > 3k_{inj}$, the transfer rate of the enstrophy is observed to be constant, and the energy spectrum shows a power-law scaling $E(k) \propto k^{-\alpha}$ in the inertial range $k_{inj} \leq k \leq k_d$ with α larger than the theoretical prediction of 3. This discrepancy is attributed to the inhibition of the cascade process reflecting the effect of the long persistence of coherent patches.

ACKNOWLEDGMENTS

This research was supported by the Grant-in-Aid for Scientific Research (B) 17340173 of JSPS and partly by the collaborative program of NIFS.

REFERENCES

1. C. F. Driscoll, D. Z. Jin, D. A. Schecter and D. H. E. Dubin, Physica C **369**, 21 (2002).
2. A. J. Peurrung and J. Fajans, Phys. Fluids A **5**, 493 (1993).
3. Y. Kiwamoto, K. Ito, A. Sanpei and A. Mohri, Phys. Rev. Lett. **85**, 3173 (2000).
4. K. S. Fine, A. C. Cass, W. G. Flynn and C. F. Driscoll, Phys. Rev. Lett. **75**, 3277 (1995).
5. Y. Kiwamoto, N. Hashizume, Y. Soga, J. Aoki, and Y. Kawai, Phys. Rev. Lett. **99**, 115002 (2007).
6. Y. Kawai, Y. Kiwamoto, K. Ito, A. Sanpei, Y. Soga, J. Aoki and K. Itoh, J. Phys. Soc. Jpn. **75**, 104502 (2006).
7. Y. Kawai, Y. Kiwamoto, Y. Soga and J. Aoki, Phys. Rev. E **75**, 066404 (2007).
8. R. H. Kraichnan, Phys. Fluids **10**, 1417 (1967). G. K. Batchelor, Phys. Fluids **12** (Suppl.II), 233 (1969).
9. Y. Kawai and Y. Kiwamoto, Phys. Rev. E (to be published).
10. P. Tabeling, Physics Reports **362**, 1 (2002).
11. M. A. Rutgers, Phys. Rev. Lett. **81**, 2244 (1998).
12. D. H. E. Dubin and T. M. O'Neil, Phys. Plasmas. **5**, 1305 (1998).
13. J. C. McWilliams, J. Fluid Mech. **146**, 21 (1984). P. Santangelo, R. Benzi and B. Legras, Phys. Fluids A **1**, 1027 (1989).

Collisional Damping Of Plasma Waves On A Pure Electron Plasma Column

M. W. Anderson and T. M. O'Neil

Department of Physics, University of California at San Diego, 9500 Gilman Drive, La Jolla, California 92093, USA

Abstract. The collisional damping of electron plasma waves (or, more precisely, Trivelpiece-Gould waves) on a pure electron plasma column is discussed. The damping in a pure electron plasma differs from that in a neutral plasma, since there are no ions to provide collisional drag. A dispersion equation for the complex wave frequency is derived from Poisson's equation and the drift-kinetic equation with the Dougherty collision operator—a Fokker-Planck operator that conserves particle number, momentum, and energy yet is analytically tractable. In the limit of weak collisionality, for phase velocity comparable to the thermal velocity, Landau damping is recovered. For larger phase velocity, where Landau damping is negligible, the dispersion equation can be solved analytically, yielding the complex frequency $\omega = (k_z \omega_p / k)[1 + (3/2)(k\lambda_D)^2(1 + 10i\alpha/9)(1 + 2i\alpha)^{-1}]$, where ω_p is the plasma frequency, k_z is the axial wavenumber, k is the total wavenumber, λ_D is the Debye length, ν is the collision frequency, and $\alpha \equiv \nu k / \omega_p k_z$. This expression spans from the weakly collisional regime ($\alpha \ll 1$) to the strongly collisional regime ($\alpha \gg 1$), matching onto fluid results in the latter limit. Note that in the weakly collisional regime the damping rate is given by $\mathrm{Im}(\omega) \cong -4\nu(k\lambda_D)^2/3$, which is suppressed from the damping rate in a neutral plasma [$\mathrm{Im}(\omega) \cong -\nu/2$] by the factor $(k\lambda_D)^2 \ll 1$; this suppression reflects the conservation of electron momentum in the pure electron plasma. In the limit of strong collisionality, the damping is enhanced by cross-field transport resulting from long-range collisions. These collisions are neglected in the kinetic treatment, but their contribution to the damping is estimated from fluid theory.

INTRODUCTION

The collisionless damping of plasma waves has been studied extensively in both neutral and nonneutral plasmas [1,2], but for waves with sufficiently large phase velocity, collisional effects provide the dominant contribution to the damping. The collisional damping of plasma waves in neutral plasmas was analyzed first by Bohm and Gross [3] and later by Lenard and Bernstein [4] in the 1950's; however, the collisional damping of waves in a pure electron plasma is fundamentally different due to the absence of ions [5].

In a neutral plasma, immobile ions exert collisional drag on the electrons, damping their oscillatory motion. Bohm and Gross determined that for a plasma wave with large phase velocity [that is, for $\mathrm{Re}(\omega)/k_z \gg v_{th}$, where ω is the wave frequency, k_z is the wave number, and v_{th} is the thermal velocity], the damping rate is given by $\mathrm{Im}(\omega) \cong -\nu_{ei}/2$, where ν_{ei} is the electron-ion slowing-down rate. Lenard and Bernstein arrived at the same result using a simple Fokker-Planck collision operator whose action on the electron distribution, f, is given by

$$C_{LB}(f) = v_{ei} \frac{\partial}{\partial \bar{v}} \cdot \left[v_{th}^2 \frac{\partial f}{\partial \bar{v}} + \bar{v} f \right], \qquad (1)$$

where v_{th}^2 is a constant. The Lenard-Bernstein operator does not conserve electron momentum or energy, reflecting the absorption of these quantities by the ion species.

Here we investigate the collisional damping of plasma waves on a pure electron plasma column. In this case there are no ions to absorb the momentum and energy of the electrons, so we expect the damping to be suppressed. In order to capture this distinction, we require a collision operator that conserves electron momentum and energy; to this end, we employ the Dougherty collision operator, defined by [6]

$$C_D(f) = v(n,T) \frac{\partial}{\partial \bar{v}} \cdot \left[\frac{T[f]}{m} \frac{\partial f}{\partial \bar{v}} + (\bar{v} - \bar{V}[f]) f \right], \qquad (2)$$

where $n[f] = \int d\bar{v} f$, $\bar{V}[f] = n^{-1} \int d\bar{v} \bar{v} f$, $T[f] = (3n)^{-1} \int d\bar{v} m (\bar{v} - \bar{V})^2 f$, $v(n,T) = ne^4 \log(r_c/b) m^{-1/2} T^{-3/2}$, r_c is the cyclotron radius, and b is the classical distance of closest approach. For simplicity, we consider an infinite, cylindrical plasma column immersed in a uniform axial magnetic field and enclosed in a conducting cylinder, and we limit the discussion to waves symmetric about the axis of the column. A more detailed analysis can be found in [5].

COLLISIONAL DRIFT-POISSON SYSTEM

The starting point of our analysis is the collisional drift-Poisson system of equations for the distribution of guiding centers, $f(r,z,v_z,v_\perp^2,t)$, and the potential, $\varphi(r,z,t)$:

$$\frac{\partial f}{\partial t} + v_z \frac{\partial f}{\partial z} + \frac{c\hat{z} \times \bar{\nabla}\varphi}{B} \cdot \bar{\nabla} f + \frac{e}{m} \frac{\partial \varphi}{\partial z} \frac{\partial f}{\partial v_z} = C_D(f) \qquad (3)$$

$$\nabla^2 \varphi = 4\pi e \int d\bar{v} f. \qquad (4)$$

In the drift approximation, the Dougherty collision term takes the form

$$C_D(f) = v(n,T) \frac{1}{v_\perp} \frac{\partial}{\partial v_\perp} v_\perp \left[\frac{T[f]}{m} \frac{\partial f}{\partial v_\perp} + v_\perp f \right] \\ + v(n,T) \frac{\partial}{\partial v_z} \left[\frac{T[f]}{m} \frac{\partial f}{\partial v_z} + (v_z - V_z[f]) f \right]. \qquad (5)$$

A steady-state solution to Eqs. (3) and (4) is given by

$$f = f_0(r, v_z, v_\perp^2) = \frac{n_0(r)}{(2\pi T_0/m)^{3/2}} e^{-m(v_z^2 + v_\perp^2)/2T_0} \tag{6}$$

$$\varphi = \varphi_0(r), \tag{7}$$

where $n_0(r)$ and $\varphi_0(r)$ are related through Poisson's equation.

For simplicity, we assume a "top-hat" density profile, given by

$$n_0(r) = \begin{cases} n_0 & \text{for } r < R_p \\ 0 & \text{for } R_p < r < R_c, \end{cases} \tag{8}$$

where R_p and R_c are the radii of the plasma column and the conducting cylinder, respectively. In the top-hat approximation, the plasma admits eigenmodes of the form [7-9]

$$f(r, z, v_z, v_\perp^2, t) = f_0(r, v_z, v_\perp^2) + \hat{\delta f}(v_z, v_\perp^2) J_0(k_\perp r) e^{i(k_z z - \omega t)} \tag{9}$$

$$\varphi(r, z, t) = \varphi_0(r) + \delta\hat{\varphi} J_0(k_\perp r) e^{i(k_z z - \omega t)} \tag{10}$$

for $r < R_p$. Here k_z is an arbitrary wavenumber and $k_\perp$ takes on one of a discrete set of values determined by k_z together with the boundary and matching conditions on φ. Substiting Eqs. (9) and (10) into Eqs. (3) and (4), discarding second-order terms, and eliminating the potential, we obtain the eigenvalue equation [5]

$$k_z v_z \hat{\delta f} + iC_D^{(1)}(\hat{\delta f}) + \frac{k_z v_z}{k^2 \lambda_D^2} \frac{f_0}{n_0} \int d\vec{v} \hat{\delta f} = \omega \hat{\delta f}, \tag{11}$$

where $C_D^{(1)}(\hat{\delta f})$ is the linear approximation to the Dougherty collision term, $k \equiv \sqrt{k_z^2 + k_\perp^2}$, and $\lambda_D \equiv \sqrt{4\pi n_0 e^2 / T_0}$. The eigenvalue of Eq. (11) with smallest imaginary part gives the frequency and damping rate of a plasma wave on the electron column.

DISPERSION EQUATION

We define the scaled variables $\Omega \equiv \omega / k_z v_{th}$, $\mu \equiv \nu / k_z v_{th}$, and $u_{z,\perp} \equiv v_{z,\perp} / v_{th}$ (here $v_{th} \equiv \sqrt{T_0/m}$), and proceed by expressing $\hat{\delta f}$ in terms of Hermite and Laguerre polynomials as [10]

$$\hat{\delta f}(u_z, u_\perp^2) = f_0 \sum_{m,n=0}^{\infty} a_{mn} \phi_{mn}(u_z, u_\perp^2), \tag{12}$$

where

$$\phi_{mn} \equiv \frac{1}{\sqrt{m!}} He_m(u_z) L_n(u_\perp^2/2). \tag{13}$$

Substituting the expansion (12) into Eq. (11) and exploiting the orthogonality of the Hermite and Laguerre polynomials, we obtain an infinite system of linear algebraic equations for the coefficients $\{a_{mn}\}$. Fortunately, for $m > 2$, the coupling is limited to "nearest neighbors" in m; one can therefore deduce the following dispersion equation for the complex frequency Ω in terms of continued fractions [5]:

$$(k\lambda_D)^2 = \frac{F_1(\Omega, \mu) F_2(\Omega, \mu) + 8\mu^2/9}{[F_1(\Omega, \mu) F_2(\Omega, \mu) + 8\mu^2/9](\Omega^2 - 1) - 2F_2(\Omega, \mu)\Omega} \tag{14}$$

where

$$F_1(\Omega, \mu) \equiv \Omega + \frac{4}{3}i\mu - \cfrac{3}{\Omega + 3i\mu - \cfrac{4}{\Omega + 4i\mu - \cdots}} \tag{15}$$

$$F_2(\Omega, \mu) \equiv \Omega + \frac{2}{3}i\mu - \cfrac{1}{\Omega + 3i\mu - \cfrac{2}{\Omega + 4i\mu - \cdots}}. \tag{16}$$

For a particular choice of μ and $k\lambda_D$, the least-damped root of Eq. (14) gives the complex frequency of a plasma wave. In the limit $\mu \to 0$, Landau damping is recovered.

ANALYTIC SOLUTION FOR LARGE PHASE VELOCITY

For $k\lambda_D \ll 1$, the least-damped root of Eq. (15) corresponds to a plasma wave with large phase velocity [Re(Ω) >> 1] and weak damping [Re(Ω) >> Im(Ω)]. In this case, the dispersion equation can be solved as an expansion in $k\lambda_D$ and μ^{-1}, and the desired root is given by

$$\Omega \cong (k\lambda_D)^{-1} \left\{ 1 + \frac{3}{2}(k\lambda_D)^2 \left[\frac{1 + 10i\mu k\lambda_D/9}{1 + 2i\mu k\lambda_D} \right] \right\}. \tag{17}$$

Restoring the units yields the more recognizable form

$$\omega \cong \omega_p \frac{k_z}{k} \left\{ 1 + \frac{3}{2}(k\lambda_D)^2 \left[\frac{1+10i\alpha/9}{1+2i\alpha} \right] \right\}, \tag{18}$$

where $\alpha \equiv \nu k/\omega_p k_z \cong \nu/\text{Re}(\omega)$ is a parameter characterizing the collisionality of the dynamics.

In the limit of weak collisionality ($\alpha \ll 1$), Eq. (18) reduces to

$$\text{Re}(\omega) \cong \frac{k_z \omega_p}{k}\left[1+\frac{3}{2}(k\lambda_D)^2\right], \tag{19}$$

$$\text{Im}(\omega) \cong -\frac{4}{3}\nu\left(\frac{k_z v_{th}}{\omega}\right)^2 \cong -\frac{4}{3}\nu(k\lambda_D)^2. \tag{20}$$

Equation (19) is the well-known result from collisionless theory for the frequency of an electron plasma wave—or, more precisely, a Trivelpiece-Gould wave—on a pure electron plasma column [7-9]. The collisional damping rate, given by Eq. (20), is reduced from the rate for a plasma wave in a neutral plasma by the factor $(k_z v_{th}/\omega)^2 \ll 1$. This reduction is a reminder that the dominant damping mechanism in a neutral plasma—electron-ion friction—is not available in a pure electron plasma.

In the limit of strong collisionality ($\alpha \gg 1$), Eq. (18) yields

$$\text{Re}(\omega) \cong \frac{k_z \omega_p}{k}\left[1+\frac{5}{6}k^2\lambda_D^2\right], \tag{21}$$

$$\text{Im}(\omega) \cong -\frac{1}{3}\left(\frac{v_{th}^2}{\nu}\right)k_z^2. \tag{22}$$

[Note: In discussing the strongly collisional limit, we are limited to the consideration of waves for which $k_z \alpha \ll k$; this inequality ensures that the plasma is weakly correlated (i.e., $\nu \ll \omega_p$) despite the fact that the wave dynamics is strongly collisional.] Equations (21) and (22) match results from a fluid calculation that assumes the "Dougherty parallel viscosity", $\eta^{(D)} \equiv mn_0 v_{th}^2/2\nu$, corresponding to the Dougherty collision operator, and neglects cross-field transport [5,11]. The fluid calculation illustrates that the damping is due to bulk viscosity.

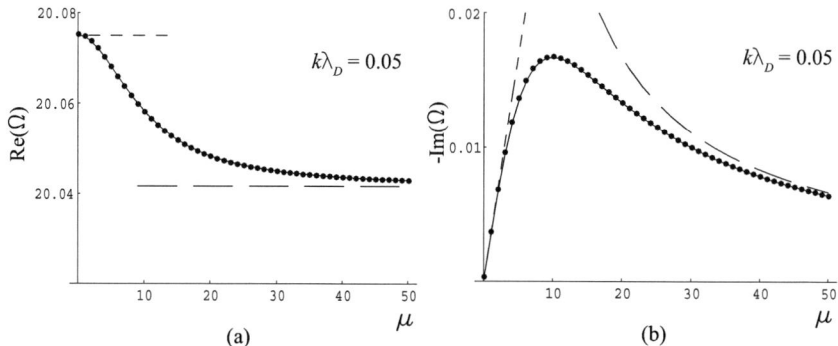

FIGURE 1. Real (a) and imaginary (b) parts of the analytic approximation (17) for the least-damped root of Eq. (14), plotted (as solid curve) versus μ, for $k\lambda_D = 0.05$. The solid circles give the numerical solution to Eq. (14). The short-dashed curves give the asymptotic forms (19) and (20), which are valid for $\mu \ll \Omega$, while the long-dashed curves give the asymptotic forms (21) and (22), valid for $\mu \gg \Omega$.

If the cross-field transport terms are retained in the fluid equations, the resulting damping rate has an extra term that does not appear in Eq. (22), which is proportional to the cross-field viscosity, $\eta_\perp$:

$$\Delta \text{Im}(\omega) = -k_\perp^2 \eta_\perp / 2. \quad (23)$$

The drift-kinetic treatment presented above omits all cross-field transport and thus misses this contribution. In the parameter regime typical of experiments on pure electron plasmas, cross-field transport is dominated by long-range collisions [12-14] with impact parameter $\rho \sim \lambda_D \gg r_c$, giving rise to the viscosity $\eta_\perp \cong 0.6 n v_{th} b^2 \lambda_D^2$. Equation (23) then yields

$$\Delta \text{Im}(\omega) \cong -0.3 n v_{th} b^2 k_\perp^2 \lambda_D^2 = -0.3 \frac{v k_\perp^2 \lambda_D^2}{\log(r_c/b)}. \quad (24)$$

Evidently, for sufficiently strong collisionality, the contribution to the damping from cross-field viscosity [Eq. (24)] supersedes that from bulk viscosity [Eq. (22)].

ACKNOWLEDGEMENTS

The authors thank Professor D. H. E. Dubin for useful discussions. This work was supported by NSF grant PHY-0354979.

REFERENCES

1. J. H. Malmberg and C. B. Wharton, Phys. Rev. Lett. **13**, 184 (1964).

2. J. R. Danielson, F. Anderegg, and C. F. Driscoll, Phys. Rev. Lett. **92**, 245003 (2004).
3. D. Bohm and E. P. Gross, Phys. Rev. **75**, 1851 (1949).
4. A. Lenard and I. B. Bernstein, Phys. Rev. **112**, 1456 (1958).
5. M. W. Anderson and T. M. O'Neil, Phys. Plasmas **14**, 112110 (2007).
6. J. P. Dougherty, Phys. Fluids **7**, 1788 (1964).
7. A. W. Trivelpiece and R. W. Gould, J. Appl. Phys. **30**, 1784 (1959).
8. S. A. Prasad and T. M. O'Neil, Phys. Fluids **26**, 665 (1983).
9. R. C. Davidson, *Theory of Nonneutral Plasmas* (Imperial College Press and World Scientific Publishing, New York, 2001), p. 251.
10. C. S. Ng, A. Bhattacharjee, and F. Skiff, Phys. Rev. Lett. **83**, 1974 (1999).
11. M. W. Anderson and T. M. O'Neil, Phys. Plasmas **14**, 052103 (2007).
12. D. H. E. Dubin, Phys. Plasmas **5**, 1688 (1998).
13. T. M. O'Neil, Phys. Rev. Lett. **55**, 943 (1985).
14. D. H. E. Dubin and T. M. O'Neil, Phys. Rev. Lett. **78**, 3868 (1997).

Theory and Simulation of Neoclassical Transport Processes, with Local Trapping

Daniel H. E. Dubin

Department of Physics, University of California at San Diego, La Jolla, CA USA 92093-0319

Abstract.
Neoclassical transport is studied using idealized simulations that follow guiding centers in given fields, neglecting collective effects on the plasma evolution, but including collisions at rate ν. For simplicity the magnetic field is assumed to be uniform; transport is due to asymmetries in applied electrostatic fields. Also, the Fokker-Planck equation describing the particle distribution is solved, and the predicted transport is found to agree with the simulations. Banana, plateau, and fluid regimes are identified and observed in the simulations. When separate trapped particle populations are created by application of an axisymmetric squeeze potential, enhanced transport regimes are observed, scaling as $\sqrt{\nu}$ when $\nu < \omega_0 < \omega_b$ and as $1/\nu$ when $\omega_0 < \nu < \omega_b$ where ω_0 and ω_b are the rotation and axial bounce frequencies, respectively. These regimes are similar to those predicted for neoclassical transport in stellarators.

Keywords: transport, neoclassical, nonneutral plasma
PACS: 52.25.Fi, 52.56.Ff, 52.27.Jt

Irreversible processes driven by the interaction of a plasma with static electric and/or magnetic fields are of central importance in plasma theory and experiment. For example, in the theory of neoclassical transport, a magnetically confined plasma interacts with static electric and/or magnetic field asymmetries, causing irreversible flows of particles, momentum, and energy across the magnetic field [1, 2, 3, 4]. While neoclassical theory is well developed, experiments have never fully tested the theory. In neutral plasma experiments, early work on quiescent discharges was broadly consistent with neoclassical theory [5]; but in many experiments neoclassical transport is masked by anomalous transport caused by nonlinear saturation of collective plasma instabilities. In non-neutral plasma experiments, where such instabilities are absent, detailed measurements of transport over the course of several decades have still failed to make close contact with neoclassical theory [6, 7, 8, 9]. Interpretation of experimental results is often complicated by the interplay of multiple effects, even in the simplest experimental design.

In order to clarify the reasons behind observed discrepancies between neoclassical theory and experiments in nonneutral plasmas, the theory has recently been recast to consider electrostatic field errors of the type typically encountered in nonneutral experiments, and simplified for cylindrical geometry and uniform magnetic field [10]. In this recent work, specific examples were analyzed theoretically and compared to simulations that measure the transport. To further simplify the theory and simulations, plasma shielding effects on the asymmetry potentials were neglected. A local approximation to the kinetic equation, valid in the transport limit where the field error potential is much smaller than the plasma temperature, allowed the determination of local transport coefficients that link dissipative cross- field particle, momentum and energy fluxes to plasma rotation, parallel velocity, and temperature and velocity gradients. In particular, temperature-gradient-driven particle flux can be important if the gradient is sufficiently large. In non-neutral plasma experiments such large temperature gradients often develop naturally during the transport process itself, as the plasma expands radially and converts some of its electrostatic potential energy into heat.

In each example considered, the transport simulations agreed with the theory. However, such detailed comparisons require rather precise knowledge of the plasma potential, both of the zeroth-order equilibrium and the asymmetry. For instance it was observed that if, in the zeroth order equilibrium, there exists separate trapped particle populations caused by an azimuthally symmetric squeeze potential, and if the rotation frequency is small compared to the bounce frequency, the transport is strongly modified from the banana and plateau regime predictions. New $1/\nu$ and $\sqrt{\nu}$ regimes were found similar to those predicted in neoclassical transport theory for toroidal plasmas [11, 8]. Even a small population of such trapped particles completely changes the magnitude and scaling of the transport from theory predictions in the absence of trapping. Furthermore, theory presently under development suggests that small θ-asymmetries in the squeeze potential can further increase the transport due to chaotic separatrix crossing.

This paper briefly describes the theory and simulations presented in Ref. [10]. The transport theory assumes that a guiding center description of the particle motion is sufficient. In the absence of collisions, the guiding center position is described by cylindrical coordinates (r, θ, z), and only the momentum parallel to the magnetic field p_z is followed; the kinetic energy perpendicular to the field is an adiabatic invariant and is not required. Equations of motion for the guiding center are of Hamiltonian form:

$$\frac{d\theta}{dt} = \frac{\partial H}{\partial p_\theta}, \quad \frac{dp_\theta}{dt} = -\frac{\partial H}{\partial \theta},$$
$$\frac{dz}{dt} = \frac{\partial H}{\partial p_z}, \quad \frac{dp_z}{dt} = -\frac{\partial H}{\partial z} \quad (1)$$

where we follow

$$p_\theta = eBr^2/2c \quad (2)$$

rather than r because p_θ is canonically conjugate to θ, and where the Hamiltonian H is given by

$$H(\theta, p_\theta, z, p_z) = \frac{p_z^2}{2m} + \phi(\theta, p_\theta, z), \quad (3)$$

where ϕ is the electrostatic potential energy. We assume that this potential is time independent and of the form

$$\phi(p_z, \theta, z) = \phi_0(p_\theta, z) + \delta\phi(p_\theta, \theta, z) \quad (4)$$

where ϕ_0 is the equilibrium potential of the plasma, including cylindrically-symmetric external confinement fields, and $\delta\phi$ is an applied asymmetry potential that is responsible for the transport. The theory assumes that $\delta\phi \ll T$, where T is the plasma temperature, in order that the plasma expansion is a slow transport process.

The single particle orbits described by Eq. (1) are by themselves insufficient to describe neoclassical transport processes; the potentials in typical experiments are such that these orbits remain confined. Collisions are needed to describe plasma loss, and are added to the theory by way of the Fokker-Planck equation for the particle distribution function $f(\theta, p_\theta, z, p_z, t)$:

$$\frac{\partial f}{\partial t} + \dot{\theta}\frac{\partial f}{\partial \theta} + \dot{p}_\theta\frac{\partial f}{\partial p_\theta} + \dot{z}\frac{\partial f}{\partial z} + \dot{p}_z\frac{\partial f}{\partial p_z} = \hat{C}f \quad (5)$$

where here we assume a collision operator of the form

$$\hat{C}f = D\frac{\partial}{\partial p_z}\left(\frac{\partial f}{\partial p_z} + \frac{p_z - mV_b}{mT_b}f\right) \quad (6)$$

and where D is the diffusion coefficient for parallel momentum, V_b is the parallel velocity of a background species with which the plasma is colliding (usually taken to be zero), and T_b is the temperature of the background. Other forms of the collision operator could be used (see Ref. [10] for some examples), but Eq. (6) is particularly easy to simulate.

Equation (5) can be solved in the limit $\delta\phi \ll T$ to obtain expressions for the fluxes of particles, energy, and momentum across the magnetic field due to the asymmetry. The fluxes are linearly related to V_b, to gradients (if any) in V_b and T_b, and to the fluid plasma rotation frequency ω_r, defined as

$$\omega_r = -\frac{\partial\bar{\phi}}{\partial p_\theta} - \frac{T_b}{\bar{n}}\frac{\partial\bar{n}}{\partial p_\theta}, \quad (7)$$

where $\bar{\phi}$ is the θ and z-averaged potential (weighted by the plasma density) and $\bar{n}$ is the θ and z-averaged density. The first term is the average $\mathbf{E} \times \mathbf{B}$ drift and the second term is related to the plasma diamagnetic drift; here we note that when $\delta\phi \ll T$, T_b is nearly the same as T because plasma heating due to the asymmetry is weak, so we could replace T_b by T in Eq. (7) [this was done in Ref. [10] in the sections of the paper that employed the collision operator given by Eq. (6)]. In particular, radial particle flux Γ_r is linearly related to rotation via a transport coefficient μ_{11}:

$$\Gamma_r = \frac{c}{eBr}\mu_{11}\omega_r = -\left(\frac{c}{eBr}\right)^2\mu_{11}\left(\frac{\partial\bar{\phi}}{\partial r} + \frac{T_b}{\bar{n}}\frac{\partial\bar{n}}{\partial r}\right) \quad (8)$$

where we have used Eqs. (3) and (7) to connect the flux to the mobility (the first term) and diffusive (the second term) fluxes. Many (15) other transport coefficients occur in the theory but will not be discussed here.

Solutions for μ_{11} were compared to particle simulations in various cases. The simulations directly measured the diffusion and mobility fluxes to provide independent measures of μ_{11}. In these simulations the Hamiltonian equations of motion given by Eq. (1) are numerically integrated forward in time, but the parallel force law is modified in order to include a nonconservative collisional drag term,

$$dp_z/dt = -\partial H/\partial z - \nu\, p_z \quad (9)$$

where ν is the collision frequency, related to the momentum diffusion coefficient D in Eq. (6) by $D = mT_b\nu$. The equations of motion are integrated numerically using a

4th order Runge-Kutta method with constant time stepsize Δt. After every time step, a random momentum is added to p_z taken from the range $[-p_0, p_0]$ in order to simulate the effect of random forcing due to the collisions. The theory of Brownian motion then implies a relation between the background temperature and the simulation parameters, assuming that $\nu \Delta t \ll 1$:

$$T_b = \frac{p_0^2}{6m\nu\Delta t}. \qquad (10)$$

The simulations follow $N \gg 1$ particles starting at the same radius with random z, p_z and θ taken from a Boltzmann distribution $\exp(-H/T_b)$. The following two quantities are evaluated in order to measure mobility and diffusion:

$$\langle \delta r \rangle = \sum_{i=1}^{N} [r_i(t) - r_i(0)] \qquad (11)$$

and

$$\langle \delta r^2 \rangle = \sum_{i=1}^{N} [r_i(t) - r_i(0)]^2. \qquad (12)$$

The radial mobility coefficient $\mu_r = (c/eBr)^2 \mu_{11}$ is related to the rate of change of the mean radial position change through

$$\bar{n}\frac{d}{dt}\langle \delta r \rangle = \mu_r \bar{E}_r \qquad (13)$$

and the radial diffusion coefficient $D_r = (c/eBr)^2 (T_b/\bar{n}) \mu_{11}$ is related to the mean square change of radial position:

$$\frac{d}{dt}\left[\langle \delta r^2 \rangle - \langle \delta r \rangle^2\right] = 2D_r. \qquad (14)$$

An example of such an evaluation of μ_{11} is shown in Fig. 1. The Hamiltonian used here is

$$H = -\omega_0 p_\theta + \frac{p_z^2}{2m} + \varepsilon \cos(\ell\theta + kz) \qquad (15)$$

where ω_0 is the $\mathbf{E} \times \mathbf{B}$ rotation frequency and the field asymmetry is taken to be a single wave with amplitude ε. In this example, $\omega_0 = k\bar{v}$ so the mobility flux is relatively easy to measure; in other cases with smaller values of ω_0 it is easier to measure diffusion in the simulations. Measuring the mobility and diffusive fluxes for different values of the collision frequency allow one to map out the standard banana, plateau and fluid regimes, as shown in Fig. 2. Also shown in the figure are the theory predictions for the transport in these regimes. One can see that the simulations follow the theory closely.

The scaling of the transport with experimental parameters in each regime can be understood from fairly simple arguments. For large collision frequencies, $\nu > k\bar{v}$, where $\bar{v} \equiv \sqrt{T/m}$, an analysis based on fluid equations provides the transport coefficients [10]. In this fluid regime, radial particle transport is primarily caused by dissipation associated with compression and expansion of the plasma as it rotates through the field error. Temperature and velocity gradients can also lead to irreversible fluxes of particles, energy and momentum as the field error transports particles across the magnetic field.

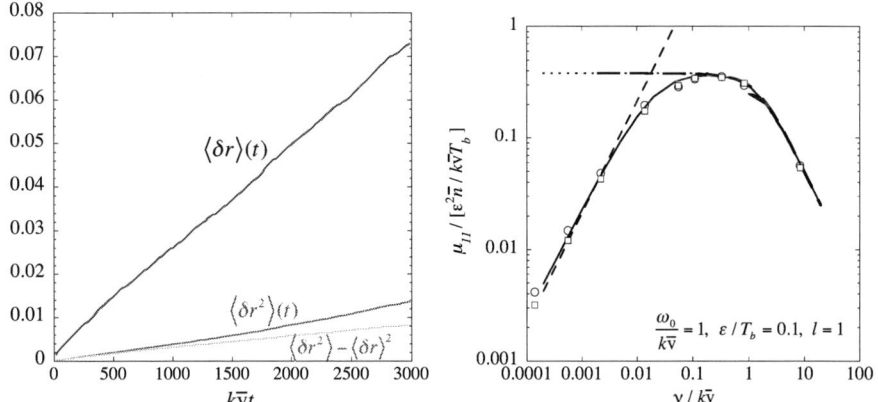

FIGURE 1. Mean and mean square change in radius for a simulation of $N = 3000$ particles following Hamiltonian (15) with $\omega_0/k\bar{v} = 1$, $\varepsilon/T_b = 0.1$, $\ell = 1$, and $v/k\bar{v} = 0.002$. Distances are in units of k.

FIGURE 2. Transport coefficient μ_{11} versus collision frequency for Hamiltonian (15), obtained by mobility measurements (squares) and diffusion measurements (dots). Lines are theory in different limits. Solid line is a full nonlinear solution of Eq. (5) for μ_{11}; thick dashed line linearizes the solution for f in $\delta\phi$; thin dashed line is the banana limit; dotted line is the plateau limit; and dot-dashed line (barely visible) is the fluid limit.

For small collision frequencies, $v < (\varepsilon/T)^{3/2}k\bar{v}$, transport coefficients are linear functions of v [12, 13]. The scaling of the radial diffusion coefficient D_r in this "banana regime" may be understood from the following argument. Particles become trapped in the field asymmetry when they have an axial velocity v_z that satisfies

$$\frac{m}{2}(v_z - \ell\omega_0/k)^2 < \varepsilon. \tag{16}$$

The trapped particles execute axial oscillations in the field error at roughly the trapping frequency $\omega_T = \sqrt{k^2\varepsilon/m}$. In these axial trapping oscillations, the particles also drift radially, with a radial "banana orbit" width $\Delta r \sim \ell\sqrt{2\varepsilon/m}/kr\Omega_c$ where $\Omega_c = eB/mc$. This estimate follows from the product of the radial drift velocity $\ell c\varepsilon/eBr$ and the period ω_T^{-1} of the oscillation. Transport occurs as particles become collisionally detrapped and then retrapped.

The size of the step in this process is Δr. The time between steps is the time Δt required to be detrapped from the banana orbit, $\Delta t \sim \varepsilon/(vT)$ (the time needed to diffuse in energy by order ε). The radial particle diffusion coefficient D_r is therefore roughly

$$D_r \sim f\frac{\Delta r^2}{\Delta t} \tag{17}$$

where f is the fraction of particles that take part in the banana orbits, of order $f \sim e^{-\ell^2\omega_0^2/2k^2\bar{v}^2}\sqrt{\varepsilon/T}$ for a Maxwellian distribution. Putting these estimates together yields

$$D_r \sim v\sqrt{\frac{\varepsilon}{T}}\frac{\ell^2\bar{v}^2}{k^2r^2\Omega_c^2}e^{-\ell^2\omega_0^2/2k^2\bar{v}^2} \tag{18}$$

in the banana regime. A more rigorous derivation [10] yields Eq. (18) with a numerical coefficient of 1.1. This is plotted in Fig. 2 as a thin dashed line. This banana regime estimate is sensible only when the particles are able to execute a full trapping oscillation before they are collisionally detrapped: this requires $\omega_T \Delta t \gtrsim 1$, which implies

$$(\varepsilon/T)^{3/2} k\bar{v} > \nu \tag{19}$$

for the banana regime.

For $k\bar{v}(\varepsilon/T)^{3/2} < \nu < k\bar{v}$, the transport is in the plateau regime. Trapped particles no longer complete an entire banana orbit, so the size of the radial step is reduced to $\Delta r \omega_T \Delta t$. The diffusion coefficient is now given by

$$D_r \sim f_v \frac{(\Delta r \omega_T \Delta t)^2}{\Delta t} \tag{20}$$

where f_v is the fraction of particles in resonance with the error,

$$f_v \sim e^{-\ell^2 \omega_0^2 / 2k^2 \bar{v}^2} / (k\bar{v}\Delta t). \tag{21}$$

This estimate yields the plateau regime diffusion coefficient

$$D_r \sim \left(\frac{\varepsilon}{T}\right)^2 \frac{\ell^2 \bar{v}^3}{kr^2 \Omega_c^2} e^{-\ell^2 \omega_0^2 / 2k^2 \bar{v}^2}. \tag{22}$$

A rigorous derivation of plateau regime transport for Hamiltonian (15) yields Eq. (22) with coefficient $\sqrt{\pi/8}$. This is plotted in Fig. 2 as a dotted line, and in Fig. 4 as a dashed line.

In this example, the particle motion is unconstrained in the z direction. More typical of many nonneutral plasma experiments are finite length plasmas confined in z by applied potentials. In the next example the transport in such a plasma is simulated using the following Hamiltonian:

$$H = \frac{p_z^2}{2m} - \omega_0 p_\theta + T_b \left(\frac{z}{L}\right)^8 + \varepsilon \cos\theta \sin kz \tag{23}$$

where L is the plasma length, taken to be $kL = 4.217$.

The particle transport is measured in the same way as before, and shows similar behavior compared to the previous example, with banana, plateau, and fluid regimes apparent, and theory and simulations are again in agreement (see Fig. 3). It is important to note however that the transport in the banana and plateau regimes is a fairly sensitive function of the plasma parameters. For instance, a plot of the plateau regime value for μ_{11} is displayed in Fig. 4, evaluated for both the infinite length example of Eq. (15) and the finite length example of Eq. (23). Although the field asymmetries are very similar in the two examples, the transport displays considerably more structure in the second case due to the complex interplay between different bounce harmonics. In interpreting experiments, this shows that the plasma parameters such as rotation frequency and axial potential must be very well characterized in order for theory to be able to make useful predictions of the transport.

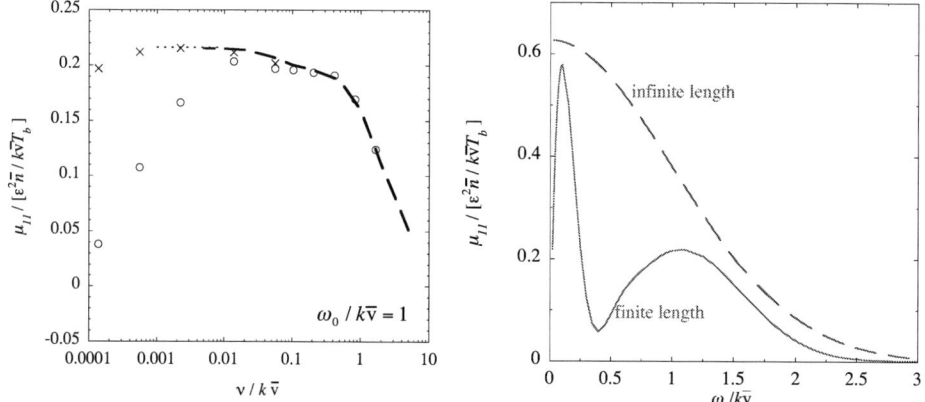

FIGURE 3. Transport coefficient μ_{11} versus collision frequency for finite length Hamiltonian (23), using simulations of diffusion for $\varepsilon/T_b = 0.02$ (dots) and $\varepsilon/T_b = 0.002$ (crosses). Thick dashed line is the theory solution of Eq. (5) for μ_{11} when f is linearized in $\delta\phi$; the dotted line is the plateau regime limit.

FIGURE 4. Plateau regime limits versus rotation frequency for the infinite length case of Hamiltonian (15) (dashed) and the finite length case of Hamiltonian (23) (solid).

Sensitive dependence of the transport on the applied potentials is also exemplified in the next case considered, in which the addition of an axisymmetric squeeze potential that creates trapped particle populations that are separated by a separatrix is found to completely change the magnitude and scaling of the transport, even when the fraction of trapped particles is small. The Hamiltonian in this case is taken to be

$$H = \frac{p_z^2}{2m} - \omega_0 p_\theta + T_b \left(\frac{z}{L}\right)^8 + V_{sq} e^{-50(kz)^4} + \varepsilon \cos\theta \sin kz \quad (24)$$

where V_{sq} is the magnitude of the applied squeeze potential. When the plasma rotation frequency ω_0 is small compared to $k\bar{v}$, two new transport regimes occur, scaling with collision frequency as $1/\nu$ and $\sqrt{\nu}$.

The $1/\nu$ regime occurs when $\omega_0 < \nu < k\bar{v}$, and the transport can be understood from the following scaling argument: Since the field error potential happens to be an odd function of z, low-energy particles trapped in the $z < 0$ well created by the squeeze potential experience the opposite field error potential from those trapped in the $z > 0$ well. As a result, the $\mathbf{E} \times \mathbf{B}$ drift orbits of particles in these two wells are displaced relative to one another, and relative to untrapped particles. The magnitude Δr of the radial displacement is of order

$$\Delta r \sim \frac{\varepsilon}{m\Omega_c \omega_0 r}. \quad (25)$$

As particles wander in energy due to collisions they become detrapped and then retrapped on a timescale of order ν^{-1}, assuming that V_s is of order T_b. Since particles only complete a fraction of a drift orbit in this time, the magnitude of the radial drift step

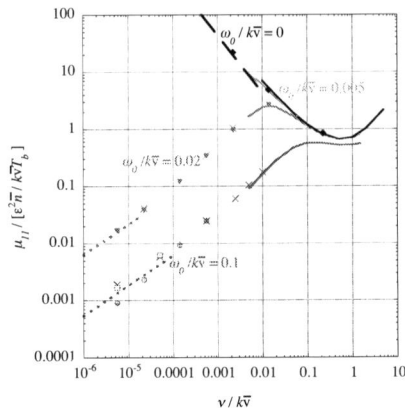

FIGURE 5. Transport coefficient μ_{11} versus collision frequency for a plasma with added squeeze given by Hamiltonian (24), with $V_{sq}/T = 0.5$, at four rotation frequencies. Symbols display simulation results. The dashed line is the $1/\nu$ regime result of Eq. (27); the dotted lines are the $\sqrt{\nu}$ regime results. Solid lines are solutions of Eq. (5) for μ_{11}, linearizing f in $\delta\phi$.

that they make is of order $\Delta r \omega_0/\nu$. The radial diffusion coefficient is then

$$D_r \sim \nu \left(\frac{\Delta r \omega_0}{\nu}\right)^2 \sim \frac{1}{\nu}\left(\frac{\varepsilon}{m\Omega_c r}\right)^2. \qquad (26)$$

In the $1/\nu$ regime, the particle diffusion increases as ν decreases,[14, 15] up to the point where $\nu < \omega_0$.

A more rigorous derivation of the transport in the $1/\nu$ regime yields, for Hamiltonian (24) with squeeze potential $V_{sq} = 0.5T_b$,

$$\mu_{11} = 0.0444 \frac{\bar{n}\varepsilon^2}{\nu T_b}. \qquad (27)$$

This result is displayed in Fig. 5 and matches the simulations when ω_0 is sufficiently small.

However, when ν falls below ω_0, transport begins to decrease with decreasing ν, as the transport enters the $\sqrt{\nu}$ regime. The transport in this regime can be understood in the following way. Particles trapped by the squeeze potential feel a different bounce-averaged field error than particles that are untrapped. The untrapped particles, bouncing rapidly from one end of the plasma to the other, average out the asymmetry potential so that, on average, they feel no net effect from the asymmetry. However, trapped particles on either side of the squeeze feel opposite asymmetry potentials, $\delta\bar{\phi} \sim \pm\varepsilon\cos\theta$. This difference between trapped and untrapped particles creates, in the absence of collisions, a discontinuity in the distribution function f at the separatrix. In the frame of the plasma, the discontinuity oscillates in time as the plasma rotates through the asymmetry. If collisions are now taken into account, the oscillating discontinuity is smoothed out by diffusion of particles in energy, over a boundary layer around the separatix of width

$\sqrt{TV_{sq}\nu/\omega_0}$. This sort of boundary layer is common in driven diffusion problems, where an oscillating source at frequency ω_0 creates an oscillating particle distribution that spreads from the source a distance of order $\sqrt{D/\omega_0}$, where D is the diffusion coefficient.

This oscillating distribution creates radial particle transport in the following way. Every rotation period, particles in the boundary layer diffuse back and forth across the separatrix, going from trapped to untrapped orbits. When the particles are trapped, they take a radial step of approximate magnitude given by Eq. (25). However, the sign of this step is random because particles are equally likely to be trapped on either side of the squeeze potential. The diffusion coefficient is the rate at which these random steps are taken, given in this case by the rotation frequency, multiplied by the square of the step, and finally multiplied by the fraction of particle participating, ie. the fraction of particles in the boundary layer:

$$D_r \sim \omega_0 \Delta r^2 \times \sqrt{\nu/\omega_0}\, e^{-V_{sq}/T} \sim \sqrt{\nu\omega_0}\, e^{-V_{sq}/T} \left(\frac{\varepsilon}{m\Omega_c\omega_0 r}\right)^2. \quad (28)$$

The scaling of this estimate for the radial diffusion coefficient agrees with a rigorous transport calculation shown in Fig. 5 as the dashed lines for two different values of the rotation frequency. Note that this transport remains finite even when $V_{sq} \ll T$, provided that the energy width of the boundary layer is small compared to V_{sq}, i.e. $\nu/\omega_0 < V_{sq}/T$. Thus, even small squeeze potentials can have a significant impact on the transport.

Finally, we note that the transport can be further enhanced if the squeeze potential is itself asymmetric in θ. In this case as particles near the separatix energy rotate in θ, they can become trapped and untrapped along *collisionless* orbits. This collisionless transport mechanism will be explored further in future work.

ACKNOWLEDGMENTS

This work was supported by National Science Foundation Grant No. PHY0354979 and NSF/DOE grant PHY0613740.

REFERENCES

1. F.L. Hinton and R.D. Hazeltine, Rev. Mod. Phys. **48**, 239 (1976).
2. H.E. Mynick, Phys. Plasmas **13**, 058102 (2006).
3. F.L. Hinton and M.N. Rosenbluth, Phys. Fluids **16**, 836 (1983).
4. R.D. Hazeltine, F.L. Hinton, and M.N. Rosenbluth, Phys. Fluids **16**, 1645 (1973).
5. T. Ohkawa, J.R. Gilleland and T. Tamano, Phys. Rev. Lett. **28**, 1107 (1972).
6. J.H. Malmberg and C.F. Driscoll, Phys. Rev. Lett. **44**, 654 (1980).
7. C.F. Driscoll and J.H. Malmberg, Phys. Rev. Lett. **50**, 167 (1983).
8. D.L. Eggleston and B. Carillo, Phys. Plasmas **10**, 1308 (2003).
9. A.A. Kabantsev, J.H. Yu, R.B. Lynch and C.F. Driscoll, Phys. Plasmas **10**, 1628 (2003).
10. D.H.E. Dubin, Phys. Plasmas **15**, 072112 (2008).
11. M.N. Rosenbluth, D.W. Ross, and D.P. Kostomarov, Nucl. Fusion **12**, 3 (1972).
12. A.A. Galeev and R.Z. Sagdeev, Sov. Phys. JETP **26**, 233 (1968).
13. M.N. Rosenbluth, R.D. Hazeltine and F.L. Hinton, Phys. Fluids **15**, 116 (1972).
14. H.E. Mynick, Phys. Plasmas **13**, 058102 (2006).
15. A. Gibson *et al.*, Plasma Phys. **11**, 121 (1969).

Effect of a Weakly Tilted Magnetic Field on the Equilibrium of Nonneutral Plasmas in a Malmberg-Penning Trap

M. Romé[*] and I. Kotelnikov[†]

[*]*I.N.F.N. Sezione di Milano and Dipartimento di Fisica, Università degli Studi di Milano, Via Celoria 16, I-20133 Milano, Italy*
[†]*Budker Institute of Nuclear Physics, Lavrentyev Av. 11, 630090 Novosibirsk, Russia*

Abstract. The effect of a weakly tilted magnetic field perturbations on the equilibrium of a nonneutral plasma confined in a Malmberg-Penning trap is analyzed. A constraint ("condition of current closure") is introduced, that in combination with the Poisson equation allows to select admissible plasma equilibria in the trap in the presence of a non-uniform and a non-axisymmetric magnetic field. Longitudinal plasma currents (analogous to the Pfirsch–Schlüter currents in Tokamaks) appearing in a nonneutral plasma even in the absence of magnetic drifts are explicitly computed in the case of a uniformly tilted magnetic field.

INTRODUCTION

The radial confinement of non-neutral plasmas in Malmberg-Penning (MP) traps is provided by a strong axial magnetic field. This field is assumed to be uniform in most theories that deal with plasma confinement. However, small perturbations of the magnetic field may play a crucial role in the transport of non-neutral plasmas in this kind of confinement devices [1, 2]. On the other hand, it is well known that an accurate treatment of the plasma transport requires at first an analysis of the plasma equilibrium, as it is demonstrated by established theories for quasi-neutral plasma confined, e.g., in tandem mirrors [3].

The equilibrium is referred here to a time interval shorter than the asymmetry-induced plasma expansion time τ_m. If the asymmetry is small, the latter is expected to be at least greater then the axial bounce time of the particles inside the trap, τ_b, and the plasma azimuthal rotation time, $2\pi/\omega_E$, i.e., $\tau_m \gg (\tau_b, 2\pi/\omega_E)$. In general, $\tau_m \propto \varepsilon^{-2}$, where the parameter ε characterizes the smallness on the magnetic field inhomogeneity. It can be assumed that $\varepsilon \sim \delta B/B$, where $\delta \mathbf{B}$ represents the difference of the actual magnetic field from an ideal uniform magnetic field $\mathbf{B}_* = B_* \mathbf{e}_z$ directed along the symmetry axis of the cylindrical confinement device. For the small ε values achieved in existing devices the expansion time can therefore be quite large, and for a shorter time interval, $t \ll \tau_m$, it is possible to consider a slowly evolving plasma column as being in a static equilibrium.

Systematic studies of nonaxisymmetric equilibria in a MP trap have been started in Refs. [4] and [5]. In Ref. [4] the equilibrium of a nonneutral plasma column in a weakly tilted magnetic field was simulated numerically. In Ref. [5] an electrostatic asymmetry was introduced by azimuthally sectored electrodes, and the analytical treatment was limited to the case of a cold plasma with a stepwise radial density profile. Later on, three-dimensional (3D) numerical particle-in-cell (PIC) simulations of the non-neutral plasma equilibrium with quadrupole or mirror magnetic perturbations have been reported in Ref. [6]. However, similar numerical simulations are hardly able to uncover fine-structure effects that limit plasma lifetime in existing and future facilities designed to achieve improved confinement of nonneutral plasmas (relevant, e.g., to antimatter studies [7, 8]).

In Ref. [9] the equilibrium of nonneutral plasmas on a set of nested toroidal magnetic surfaces has been recently considered. This work together with the theory of quasineutral plasma equilibria in tandem mirrors [3] allows establishing a constraint on the shape of admissible plasma equilibria. Together with Poisson's equation rewritten in flux coordinates [10], this constraint constitutes a self-consistent method for determining asymmetric equilibria of nonneutral plasmas in a MP trap.

NONNEUTRAL PLASMA EQUILIBRIA

The electric current produced by the flowing electrons confined in a MP trap produces a negligible change of the magnetic field, if the electron density n is far below the Brillouin limit [11] $n_B \equiv B^2/8\pi mc^2$ (with m the particle mass and c the speed of light) except for the case of a fast rotating nonneutral plasma equilibrium [12]. The magnetic field can then be described by a scalar magnetic potential such that

$$\mathbf{B} = B_* \nabla \zeta. \tag{1}$$

Alternatively, B can be written as

$$\mathbf{B} = B_* \rho \nabla \rho \times \nabla \vartheta, \tag{2}$$

The relations $\rho = \rho(x,y,z)$, $\vartheta = \vartheta(x,y,z)$, $\zeta = \zeta(x,y,z)$ define a system of curvilinear coordinates.

The momentum balance equation for a pure electron plasma is

$$mn\left(\frac{\partial \mathbf{v}}{\partial t} + \mathbf{v}\cdot\nabla\mathbf{v}\right) = en\left(\frac{1}{c}\mathbf{v}\times\mathbf{B} - \nabla\phi\right) - \nabla p,$$

where $e < 0$ is the particle charge, $\mathbf{v}$ the fluid velocity, ϕ the electrostatic potential and p the scalar pressure. If the electron density is far below the Brillouin limit, $n \ll n_B$, and the plasma column is in a slow rotation state [12], then the $\mathbf{v}\cdot\nabla\mathbf{v}$ term is negligible in comparison with the other terms and the equilibrium equation reduces to

$$\nabla p = en\left(\frac{1}{c}\mathbf{v}\times\mathbf{B} - \nabla\phi\right). \tag{3}$$

It follows easily that $\mathbf{B}\cdot\nabla p = -en\mathbf{B}\cdot\nabla\phi$. The electron temperature T tends to be constant along the magnetic field, $\mathbf{B}\cdot\nabla T = 0$. When this situation is reached, the electron density must have the form

$$n = N(\rho,\vartheta)\exp\left[-\frac{e\phi}{T(\rho,\vartheta)}\right] \tag{4}$$

and must also be consistent with the Poisson equation. Therefore, the fundamental equilibrium equation for a pure electron plasma is

$$\nabla^2 \phi(\rho,\vartheta,\zeta) = -4\pi e N(\rho,\vartheta) \exp\left[-\frac{e\phi(\rho,\vartheta,\zeta)}{T(\rho,\vartheta)}\right]. \tag{5}$$

This equation contains two functions of ρ and ϑ, N and T, which are subject to a constraint derived below. The dependence of N and T on ϑ makes the plasma equilibria in MP traps quite different from those obtained in toroidal devices [9], where the functions $N(\rho)$ and $T(\rho)$ are fully determined by the experimental conditions and by the plasma transport processes.

Using $p = p(\rho,\vartheta,\zeta) = NT\exp(-e\phi/T)$, from the equilibrium equation (3) it follows that

$$\mathbf{v} = v_\parallel \frac{\mathbf{B}}{B} - c\frac{\partial p}{\partial \rho} \frac{\nabla\rho \times \mathbf{B}}{enB^2} - c\frac{\partial p}{\partial \vartheta} \frac{\nabla\vartheta \times \mathbf{B}}{enB^2}. \tag{6}$$

The parallel component of plasma flow, $v_\parallel$, is left undefined but it must be consistent with the steady-state condition

$$\nabla \cdot (en\mathbf{v}) = 0. \tag{7}$$

This constraint leaves a net parallel electric current of the nonneutral plasma undetermined in a toroidal confinement configuration [9], but it leads to a closure condition for the MP trap geometry, since the current vanishes at the ends of the plasma column.

The "solvability condition" is given by

$$0 = \int_{-\infty}^{\infty} \left\{ -\frac{\partial p}{\partial \rho}\frac{\partial}{\partial \vartheta}\frac{1}{B^2} + \frac{\partial p}{\partial \vartheta}\frac{\partial}{\partial \rho}\frac{1}{B^2} + \frac{1}{B^2}\frac{\partial e\phi}{\partial \vartheta}\frac{\partial}{\partial \rho}\frac{p}{T} - \frac{1}{B^2}\frac{\partial e\phi}{\partial \rho}\frac{\partial}{\partial \vartheta}\frac{p}{T} \right\} d\zeta, \tag{8}$$

where the integration is formally extended over an infinite interval (actually it covers the interval of a magnetic field line where the plasma pressure p is nonzero). The constraint (8) interrelates two functions of ρ and ϑ, namely N and T, and, in general, it allows determining $N(\rho,\vartheta)$ if $T(\rho,\vartheta)$ is given or vice versa.

One can argue, however, that $T(\rho,\vartheta)$ is not completely independent of $N(\rho,\vartheta)$. Indeed, a differential plasma rotation would result in a fast sharpening of the temperature gradient across the plasma streamlines so that even a weak transverse thermal conductivity effectively flattens the temperature along the streamlines. Therefore one can assume that $\mathbf{v} \cdot \nabla T = 0$ in addition to the condition $\mathbf{B} \cdot \nabla T = 0$ used in the derivation of Eq. (4). Dotting Eq. (6) with ∇T one finally concludes that T depends on ρ and ϑ through the dependence of N on these coordinates, i.e.,

$$T(\rho,\vartheta) = T(N(\rho,\vartheta)). \tag{9}$$

For the sake of simplicity it is assumed below that $T = $ const. This assumption is relevant to the state of global thermal equilibrium [13, 14], which is also characterized by a rigid plasma rotation.

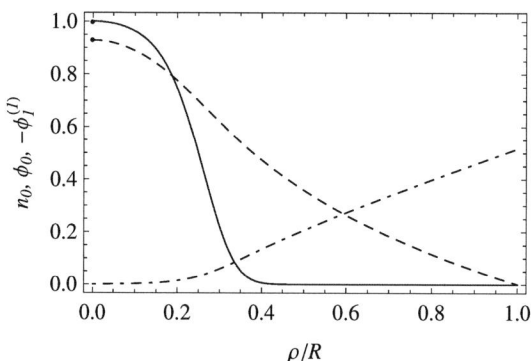

FIGURE 1. Unperturbed density n_0 (solid line), unperturbed electric potential ϕ_0 (dashed line) and radial part $\phi_1^{(1)}$ (dot-dashed line) of the perturbed potential used for calculating the Pfirsch–Schlüter currents vs. the flux radius ρ normalized over the radius R of the MP trap. The density is normalized by its maximal value n_*, and the potentials by $(T/e)(a/\lambda_D)^2$, with $\lambda_D \equiv [T/(4\pi e^2 n_*)]^{1/2}$ the Debye length. The density profiles corresponds to a global thermal equilibrium with a column radius $a/R = 0.25$ (computed at 1/2 of the maximal density), and $\lambda_D/R = 0.05$.

PFIRSCH–SCHLÜTER CURRENTS

Considering the case of a weak magnetic perturbation, $\varepsilon \ll 1$, the unknown functions ϕ and N can be sought in the form $\phi(\rho,\vartheta,\zeta) = \phi_0(\rho,\zeta) + \varepsilon\phi_1(\rho,\vartheta,\zeta)$ and $N(\rho,\vartheta) = N_0(\rho) + \varepsilon N_1(\rho,\vartheta)$. The linearized versions of Eqs. (8) and (5) can be readily solved in the region far from the plasma column ends, where the unperturbed electric potential $\phi_0 = \phi_0(\rho)$ does not depend on ζ. An example of solution is shown in Fig. 1.

Omitting the details of the calculations, in the case of a uniform magnetic field B_* tilted by a small angle α with respect to the axis of the trap, the perturbed potential can be written as

$$\phi_1(\rho,\vartheta,\zeta) = \phi_1^{(1)}(\rho)\frac{\alpha\zeta}{R}\cos\vartheta, \qquad (10)$$

while $N_1 = 0$ if $\zeta = 0$ in the midplane of the plasma column (this can be accomplished with a proper choice of the origin of the system of coordinates).

The current closure constraint (8), together with the Poisson equation (5), allows also computing the longitudinal plasma currents induced by a magnetic field perturbation in a nonneutral plasma confined in a MP trap. These currents can be thought off as an analog of the Pfirsch–Schlüter currents in Tokamaks [15] or the Stupakov currents in tandem mirrors [16, 3]. However, they appear even in the case of a uniform magnetic tilt which does not give rise to any magnetic drift, whereas both Pfirsch's–Schlüter's and Stupakov's currents originate from magnetic drifts.

The contravariant components of the perturbed electric current assume the following elegant form

$$j^i = \left\{-\frac{c}{B_*\rho}\frac{\partial p}{\partial \vartheta}, \frac{c}{B_*\rho}\frac{\partial p}{\partial \rho}, env_\parallel \frac{B}{B_*}\right\}, \qquad (11)$$

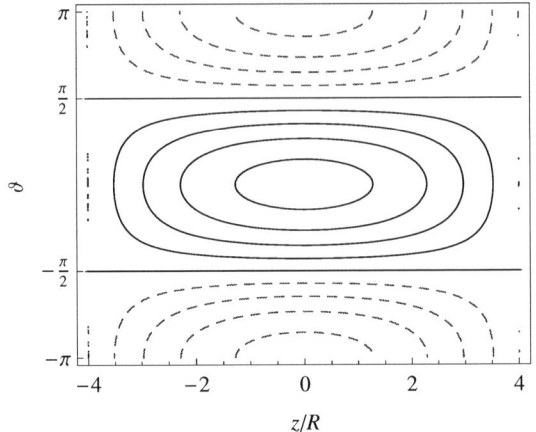

FIGURE 2. Streamlines of the perturbed current density on flux surfaces with $\rho/R = 0.24$ for a uniform magnetic field with a tilt angle $\alpha = 1°$. Solid and dashed lines correspond to a clockwise and a counterclockwise flowing current, respectively. The total length of the plasma column is $L = 4R$. Other parameters are indicated in Fig. 1.

where

$$env_\| \frac{B}{B_*} = \frac{c}{B_*} \frac{1}{\rho} \frac{\partial N_0}{\partial \rho} \int_{-\infty}^{\zeta} \frac{\partial e\phi_1}{\partial \vartheta} \exp\left(-\frac{e\phi_0}{T}\right) d\zeta. \tag{12}$$

Introducing the flux function

$$\Psi_1 \equiv \frac{cT}{B_*} \frac{\partial N_0}{\partial \rho} \int_{-\infty}^{\zeta} \frac{e\phi_1}{T} \exp\left(-\frac{e\phi_0}{T}\right) d\zeta, \tag{13}$$

the perturbed part of the electric current can be cast in the vector form

$$\mathbf{j}_1 = \nabla\rho \times \nabla\Psi_1. \tag{14}$$

This equation clearly shows that the radial current density vanishes, $j_1^1 = 0$, whereas the level contours of Ψ_1 on a surface with a given radius ρ are just the current lines.

The streamlines found for a plasma density profile corresponding to a global thermal equilibrium [13, 14] are drawn in Fig. 2. They show a similar topology at each radius and the current density is strongly suppressed close to the plasma column axis. It can be shown that these peculiar characteristics are no longer valid in the general case of a variable magnetic tilt.

CONCLUSIONS

A current closure constraint has been derived that selects a class of admissible plasma equilibria in the trap in the presence of a non-uniform and a non-axisymmetric magnetic field. In combination with Poisson's equation this constraint provides a full set of equations for determining self-consistent equilibria of nonneutral plasmas in MP traps.

The constraint of current closure has been applied to analyze plasma equilibria in the case of weakly tilted magnetic field perturbations, but the method can be straightforwardly extended to determine plasma equilibria under the effect of the magnetic perturbations of higher multipolarity (such as quadrupole or octupole fields). In particular, it can be used for stronger asymmetries such as those required to trap antihydrogen atoms in the ATHENA [7] and ATRAP experiments [8]. A suitable numerical code to determine plasma equilibria in these configurations is presently under development.

ACKNOWLEDGMENTS

Useful discussions with Prof. T. M. O'Neil, Dr. A. Kabantsev and Dr. T. Akhmetov are gratefully acknowledged.

This work has been started during a visit of I. K. to the Nonneutral Plasma Physics Group of the University of California at San Diego thanks to U.S. Civilian Research and Development Foundation Grant No RUP1-2631-NO-04. This work was also supported by the Cariplo Foundation and the Landau Network—Centro Volta.

REFERENCES

1. J. H. Malmberg and C. F. Driscoll, Phys. Rev. Lett. **44**, 654 (1980).
2. D. H. E. Dubin and T. M. O'Neil, Phys. Plasmas **5**, 1305 (1998).
3. D. D. Ryutov and G. V. Stupakov, in *Reviews of Plasma Physics*, edited by B. B. Kadomtsev (Consultants Bureau, New York, 1987), vol. 13, pp. 93-202.
4. G. W. Hart, Phys. Fluids B **3**, 2987 (1991).
5. R. Chu, J. S. Wurtele, A. Notte, A. J. Peurrung, and J. Fajans, Phys. Fluids B **5**, 2378 (1993).
6. K. Gomberoff, J. Fajans, A. Friedman, D. Grote, J.-L. Vay, and J. S. Wurtele, Phys. Plasmas **14**, 102111 (2007).
7. M. Amoretti, C. Amsler, G. Bonomi, A. Bouchta, P. Bowe, C. Carraro, C. L. Cesar, M. Charlton, M. J. T. Collier, M. Doser, V. Filippini, K. S. Fine, A. Fontana, M. C. Fujiwara, R. Funakoshi, P. Genova, J. S. Hangst, R. S. Hayano, M. H. Holzscheiter, L. V. Jørgensen, V. Lagomarsino, R. Landua, D. Lindelöf, E. Lodi Rizzini, M. Macrì, N. Madsen, G. Manuzio, M. Marchesotti, P. Montagna, H. Pruys, C. Regenfus, P. Riedler, J. Rochet, A. Rotondi, G. Rouleau, G. Testera, A. Variola, T. L. Watson, and D. P. van der Werf, Nature **419**, 456 (2002).
8. G. Gabrielse, N. S. Bowden, P. Oxley, A. Speck, C. H. Storry, J. N. Tan, M. Wessels, D. Grzonka, W. Oelert, G. Schepers, T. Sefzick, J. Walz, H. Pittner, T. W. Hänsch, and E. A. Hessels (ATRAP Collaboration), Phys. Rev. Lett. **89**, 213401 (2002).
9. T. S. Pedersen and A. H. Boozer, Phys. Rev. Lett. **88**, 205002 (2002).
10. I. Kotelnikov, M. Romé, and A. Kabantsev, Phys. Plasmas **13**, 092108 (2006).
11. L. Brillouin, Phys. Rev. **67**, 260 (1945).
12. I. Kotelnikov, M. Romé, and R. Pozzoli, Phys. Lett. A **372**, 1445 (2008).
13. T. M. O'Neil, Comments Plasma Phys. Contr. Fusion **5**, 213 (1980).
14. I. Kotelnikov, R. Pozzoli, and M. Romé, Phys. Plasmas **11**, 4396 (2000).
15. D. Pfirsch and A. Schluter, Max-Planck-Institut Report No. MPI/PA/7/62, 1962 (unpublished).
16. G. V. Stupakov, Fiz. Plazmy **5**, 871 (1987), [Sov. J. Plasma Phys. **5**, 486 (1979)].
17. T. M. O'Neil and C. F. Driscoll, Phys. Fluids **22**, 266 (1979).
18. K. Gomberoff, J. Fajans, J. Wurtele, A. Friedman, D. P. Grote, and R. H. Cohen, Phys. Plasmas **14**, 052107 (2007).

Relativistic Effects on the Radial Equilibrium of Nonneutral Plasmas

M. Romé*, I. Kotelnikov† and R. Pozzoli*

*I.N.F.N. Sezione di Milano and Dipartimento di Fisica, Università degli Studi di Milano,
Via Celoria 16, I-20133 Milano, Italy
†Budker Institute of Nuclear Physics, Lavrentyev Av. 11, 630090 Novosibirsk, Russia

Abstract. Relativistic effects on the radial equilibrium of nonneutral plasmas confined in cylindrical traps are analyzed for rigid and sheared modes of plasma rotation, both with and without the presence of a coaxial inner charged conductor. The changes with respect to the non-relativistic results are especially pronounced for the fast rotational equilibrium solutions. In particular, relativistic effects can limit the plasma outer radius. Analytical estimates of this maximum radius are found both for a rigid plasma rotation and for the case of a uniform plasma density. It is also observed that the Brillouin density limit is modified when the shielding of the external magnetic field by the current associated with the plasma rotation becomes significant.

INTRODUCTION

In the physics of nonneutral plasmas (as well as charged particle beams, accelerators, storage rings, etc.), the self-fields play a very important role for the determination of equilibrium, stability and transport properties [1, 2]. In the theoretical analysis, the transverse motion is usually considered to be non-relativistic, i.e., the self-consistent axial magnetic field produced by the azimuthal plasma current, and the relativistic modification of the centrifugal force, are neglected. This approximation is well satisfied, e.g., for charged particle beams, where the transverse velocity is usually small as compared to the mean longitudinal velocity, and the beam self-fields are typically smaller then the externally applied fields of the focusing systems; and for nonneutral plasmas confined in a Malmberg-Penning trap [3] in the slow rotational equilibrium. It is shown here that relativistic effects: i) strongly modify the equilibrium of nonneutral plasmas for the fast rotation mode even if the linear velocity of plasma rotation is small as compared to the speed of light; ii) could be easily detected experimentally; and iii) modify the Brillouin density limit [4].

In a nonneutral plasma confined in a Malmberg-Penning trap, either with or without an inner conductor (see Fig. 1), the particles azimuthally rotate due to the combined effect of the externally applied magnetic field, and the self-consistent electric field, while the mean axial velocity of the particles is zero. Within a non-relativistic cold fluid model the equilibrium of an infinitely long nonneutral plasma column with constant density, confined radially by a uniform magnetic field B_0 directed along the axis of the trap is characterized by the azimuthal rotation frequencies $\omega^\pm = -(1/2)\Omega[1 \pm (1 - 2\omega_p^2/\Omega^2)^{1/2}]$, where $\omega_p = (4\pi e^2 n/m)^{1/2}$ is the plasma frequency, with n, e and m the particle density, charge and mass, respectively, and $\Omega = eB_0/mc$ is the non-relativistic

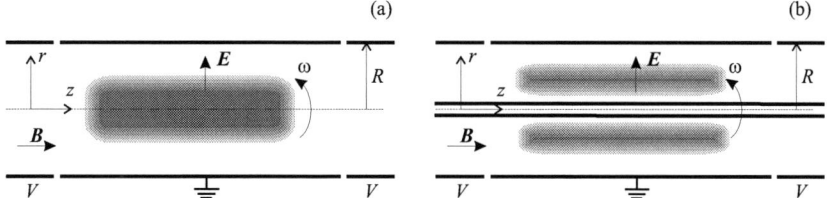

FIGURE 1. Schematics of a Malmberg-Penning trap without (a) and with an inner electrode (b). A cloud of charged particles is confined within the trap by electrostatic plugs, and rotates around the trap axis under the action of the crossed axial magnetic field **B** and the self-consistent electric field **E**. Cylindrical coordinates (r, θ, z) are used, with the z axis being the symmetry axis of the trap.

cyclotron frequency (Ω is assumed to have the same sign of the charge) [1]. The two angular frequencies ω^- and ω^+ correspond to a slow and a fast rigid rotation of the plasma column about the axis of symmetry, respectively (note that a global thermal equilibrium state with nonzero temperature also corresponds to a rigid rotation of the plasma [1]). For a low density plasma, $2\omega_p^2/\Omega^2 \ll 1$, the frequency of slow rotation is approximately equal to the electric drift (diocotron) frequency, $\omega^- \simeq -\omega_p^2/2\Omega$, while the frequency of fast rotation approaches the cyclotron frequency, $\omega^+ \simeq -(\Omega - \omega_p^2/2\Omega)$. For $2\omega_p^2/\Omega^2 = 1$, the two rotational equilibria merge, and $\omega^+ = \omega^- = -\Omega/2$. The condition $2\omega_p^2/\Omega^2 = 1$ is referred to as the Brillouin density limit: radially confined equilibria do not exist for $n > n_{B0} \equiv B_0^2/(8\pi mc^2)$ [4] in the non-relativistic limit. The two rotational equilibria have been measured experimentally [5].

RELATIVISTIC COLD FLUID RADIAL EQUILIBRIUM

Within the framework of a cold fluid model, the relativistic radial force balance equation is written as

$$-\gamma \frac{m v_\theta^2}{r} = eE_r + \frac{2e\kappa}{r} + \frac{e}{c} v_\theta B \quad (1)$$

where $\gamma = (1 - v_\theta^2/c^2)^{-1/2}$ is the relativistic factor of the fluid, κ the electric charge per unit length on the inner conductor, and the radial electric field, E_r, and the magnetic field, B, satisfy the Maxwell equations

$$\frac{1}{r}\frac{d}{dr}(rE_r) = 4\pi en, \quad \frac{dB}{dr} = -\frac{4\pi}{c} env_\theta. \quad (2)$$

Integrating with respect to the radius, the equilibrium equation can be written in terms of the azimuthal angular frequency $\omega \equiv v_\theta/r$ as

$$\gamma \omega^2 r + \frac{1}{r}\int_{r_i}^r \omega_p^2 x \, dx + \Omega \omega r + \frac{2e\kappa}{mr} - \frac{\omega r}{c^2}\int_{r_i}^r \omega_p^2 \omega x \, dx = 0, \quad (3)$$

where r_i is the inner radius of plasma ($r_i \geq r_c \geq 0$, r_c being the radius of the central conductor). In general, Eq. (3) can be solved with respect to ω for a given radial density distribution, or vice versa a solution for the density can be sought for a given radial profile of the rotation frequency.

In the special case of rigid plasma rotation, $\omega(r) = $ const, the radial equilibrium equation can be written in the form

$$v\rho^2 + \frac{v^2\rho^2}{\sqrt{1-\rho^2}} + \frac{\sigma a^2}{4} + \frac{(1-\rho^2)}{2}\int_a^\rho xN(x)\,dx = 0, \qquad (4)$$

where $\rho \equiv r|\omega|/c$, $a \equiv r_i|\omega|/c$, $v \equiv \omega/\Omega$, $N \equiv 2\omega_p^2/\Omega^2 = n/n_{B0}$, and $\sigma \equiv \kappa/(\pi r_i^2 e n_{B0})$. The solution for the density is easily determined analytically as

$$N(\rho;v,a) = -\frac{4v\left[(1-a^2)^{3/2} - va^2\right]}{(1-a^2)^{1/2}(1-\rho^2)^2} - \frac{2v^2(2+\rho^2)}{(1-\rho^2)^{5/2}}, \qquad (5)$$

where σ is uniquely determined in terms of v and a by the relation $\sigma = -4v - 4v^2/\sqrt{1-a^2}$. Physically acceptable solutions correspond to the interval of rotation frequencies $v_{\min} \equiv -2(1-a^2)^{3/2}/(2-a^2) < v < 0$. Since $\sigma > 0$ in the same range, a rigid rotation solution is allowed only when inner conductor and plasma have the same sign of charge. For given a and $N(a) \equiv N(a;v,a)$, two roots $v^\pm$ exist, corresponding to the fast and slow modes of rotation:

$$v^\pm = -\frac{(1-a^2)^{3/2}}{2-a^2}\left[1 \pm \sqrt{1 - \frac{(2-a^2)N(a)}{2(1-a^2)^{1/2}}}\right]. \qquad (6)$$

The two modes of rotation merge at $v^+ = v^- = v_{\min}/2$ where $N(a)$ attains its maximum value, $N_{\max}(a) = 2(1-a^2)^{1/2}/(2-a^2)$. In the absence of a central conductor, all formulas recover results known in the literature [1].

The density profile, Eq. (5), is defined in the interval $a < \rho < b < \rho_0$, where $b = |\omega|r_p/c$ is the normalized value of the outer plasma radius r_p, and ρ_0 denotes the limiting radius of the plasma column, defined by the condition $N(\rho_0;v,a) = 0$. In the case $a \ll 1$ and $v \to v_{\min} \simeq -(1-a^2)$, corresponding to the fast mode of rotation, the limiting radius is given by $\rho_0 \simeq \sqrt{1+v}$, and the density profile becomes approximately parabolic,

$$N(\rho;a) \simeq 4(v - v_{\min} + a^2 - \rho^2) \simeq 4(\rho_0^2 - \rho^2). \qquad (7)$$

This fast rotational equilibrium profile could be experimentally revealed, when the outer radius b of the trapped nonneutral plasma approaches the limiting value ρ_0. This condition can be easily achieved with a suitable choice of the central density and the externally applied magnetic field [7].

Due to the magnetic field shielding in the interior of the plasma column, the inequality

$$n \leq \frac{2\sqrt{1-a^2}}{2-a^2} n_B, \qquad (8)$$

holds locally for every radius and every rotation frequency, where $n_B = B^2/8\pi mc^2$ is the local value of the non-relativistic Brillouin density. The condition (8) can therefore be thought of as a relativistic generalization of the Brillouin limit for rigidly rotating annular nonneutral plasmas [8]. The absolute maximum ratio of n/n_B ($= 1$) is reached for $a = 0$ and $v = -1/2$. It is shown below that this limit may be actually overcome in the case of a sheared flow.

In the general case of sheared plasma rotation, the radial force balance equation can be written in the form

$$2v\rho^2 + \frac{2v^2\rho^2}{\sqrt{1-\rho^2v^2}} + \frac{\sigma a^2}{2} - \rho^2 v \int_a^\rho N(x)xv(x)\,dx + \int_a^\rho N(x)x\,dx = 0, \qquad (9)$$

where a different normalization of the lengths is used, so that $\rho \equiv r|\Omega|/c$. Eq. (9) must be solved for $\rho > a \equiv r_i|\Omega|/c$.

Again, for a given density profile $N(\rho)$ two solutions exist, corresponding to the fast and slow shear modes of rotation. The two equilibria can be distinguished by means of the value of the rotation frequency at the inner plasma radius, $v(a)$, which is a solution of the equation

$$v^2/\sqrt{1-v^2a^2} + v + \sigma/4 = 0. \qquad (10)$$

Two real roots, $v^+(a)$ and $v^-(a)$, of Eq. (10) exist provided that σ does not exceed the value $\sigma_m(a) = -4v_m^2(a)/\sqrt{1-v_m^2(a)a^2} - 4v_m(a)$, where $v_m(a)$ denotes the frequency at which the two roots merge, $v^+(a) = v^-(a) = v_m(a)$. The frequency of the fast and slow modes fall into the intervals $-1/a < v^+(a) < v_m(a)$ and $v_m(a) < v^-(a) < 1/a$, respectively. For a given a, no equilibria are possible if $\sigma > \sigma_m(a)$. A particular case is $\sigma = 0$, where for the fast rotational equilibrium it is $v^+(a) = -1/\sqrt{1+a^2}$, while the slow equilibrium is characterized by $v^-(a) = 0$. If $0 < \sigma < \sigma_m$ the two modes have the same sign of the frequency. For $\sigma < 0$ (corresponding to a charge on the central conductor with a sign opposite to that of the confined nonneutral plasma), the two frequencies have opposite signs, with $-1/a < v^+ < -1/\sqrt{1+a^2}$ and $0 < v^- < 1/a$. In the slow rotational equilibrium, the inner part of the annular column therefore rotates azimuthally in a direction opposite to the "natural" one, determined by the sign of the electric charge of the column and the direction of the applied magnetic field. At larger radii the effect of the column self-electric field eventually prevails, and the rotation reverts its direction.

In general, a solution of Eq. (9) may exist for all radii $\rho \geq a$ or may be defined within a limiting radius ρ_s. The analysis below is performed in a case for which a simple analytical solution exists for one of the two rotational modes. This particular solution is given by

$$v^\pm(\rho) = -1/\sqrt{2}\rho, \quad N(\rho) = 2\sqrt{2}/a, \quad B(\rho) = B_0\rho/a, \qquad (11)$$

with $\sigma = 2\sqrt{2}(1/a - 1/a^2)$, and describes an annular plasma with uniform density, rotating with constant azimuthal velocity v_θ and constant energy γ. Note that $n/n_B > 1$ can be larger than unity in the range $1 < \rho/a < 2^{3/4}/a^{1/2}$ (which exists for $a < 2^{3/2}$), demonstrating that the Brillouin limit, Eq. (8), can be overcome for a sheared flow.

From the values of $v^\pm(a)$, one readily concludes that Eq. (11) describes the fast rotation for $a < 3$ and the slow rotation for $a > 3$. This fact is in apparent contradiction with the asymptotic behaviors for $\rho \to \infty$, $v^+ \to -1/\rho$ and $v^- \to -1/\sqrt{2}\rho$ for the

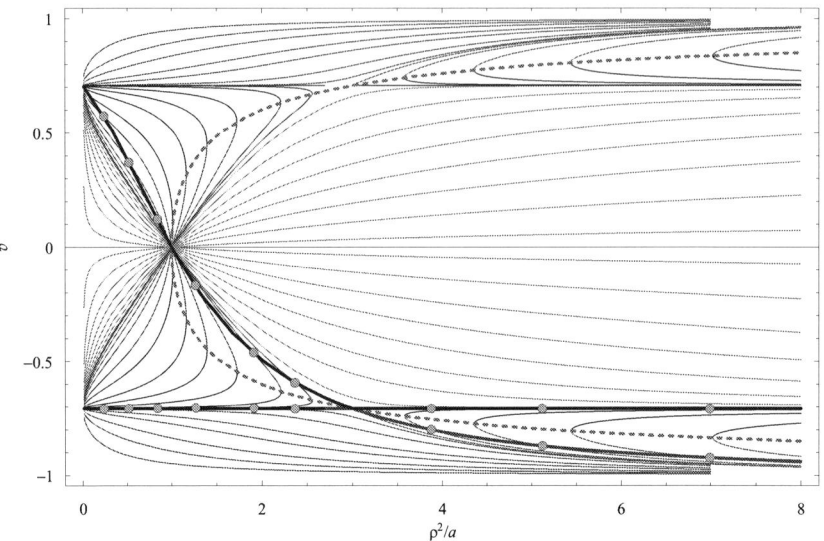

FIGURE 2. Map of frequency profiles. The solid lines are solutions of the ordinary differential equation (12). The dots denote the starting points of the trajectories which obey the original integral equation (9) for a set of values of the inner radius a.

fast and the slow mode of rotation, respectively [7], which can be determined from Eq. (9). The paradox is solved considering the behavior of the solutions of the ordinary differential equation obtained dividing the original integral equation (9) by $v(\rho) = \rho \, v(\rho)$ and differentiating the result with respect to ρ. In the present case of constant density, this equation is suitably written in terms of $v = \nu \rho$ and $x = \rho^2/a$ as

$$(v^2)' = -\frac{\left[1 - x + 2x\left(1 - v^2\right) - \sqrt{2}v^2/\left(1 - v^2\right)^{1/2}\right]v^2}{\left[1 + \sqrt{2}v^2/(1 - v^2)^{3/2} - x\right]x}, \qquad (12)$$

where the prime stands for the derivative with respect to x (note that the parameter a does not appear explicitly). Eq. (12) is singular at the point $x_s = 1 + \sqrt{2}v^2/(1 - v^2)^{3/2}$. Fig. 2 represents the numerical solution of Eq. (12) for various starting points which are considered separately for the different topological parts of the plane $(x = \rho^2/a, v = \nu\rho)$. The solutions satisfying the original integral equation (9) are the lines starting from the two points $[\rho^2/a = a, v = \nu^-(a)a]$ and $[\rho^2/a = a, v = \nu^+(a)a]$ for a given value of a and then going rightward. The starting dots corresponding to the solution $\nu = -1/\sqrt{2}\rho$ belong to the straight line $v = -1/\sqrt{2}$. It is clearly seen from Fig. 2 that this solution is a fast rotational mode for $a < 3$, which turns into the slow mode for $a > 3$. The other mode is, on the contrary, slow for $a < 3$ and fast for $a > 3$. In the case $a < 3$, the slow mode is terminated by the singular curve, where it ceases to exist. Thus this mode of sheared rotation has a limiting radius. On the contrary, in the case $a > 3$, the fast mode starts on the right of the singular curve and therefore never meets it. In this case both modes of rotation have no limiting radius.

The discussion can be extended to other sheared equilibrium solutions. In general, if a singular point ρ_s exists, one of the two modes ceases to exist at larger radii.

SUMMARY

The relativistic change of the centrifugal force and the shielding of the magnetic field by the azimuthal current due to the rotating charged column modify the equilibrium density profile of a nonneutral plasma. This modification becomes especially strong for the fast rigid rotor plasma equilibrium. In this case, there is a maximum radial extent allowed for the plasma ($\ll c/|\omega|$), and when the actual plasma radius r_p becomes comparable with this limiting radius, the density profile becomes nearly parabolic rather than stepwise as predicted by the non-relativistic theory. This fact might be verified in experiments similar to those performed by Theiss et al. more than 30 years ago [5]. Analogous limitations can be found for sheared flows. For the particular example of an annular plasma with uniform density, the outer radius cannot exceed $(3r_i c/|\Omega|)^{1/2}$, r_i being the inner radius of the plasma column. The Brillouin density limit is also modified if the shielding of the external magnetic field by the current associated with the plasma rotation is significant. The new local Brillouin density limit can be actually overcome in a sheared flow.

The presence of a mechanism limiting the plasma radial extension may indicate the existence of an instability close to the maximum allowed radius; this fact deserves further investigation.

ACKNOWLEDGMENTS

This work has been started during a visit of I. K. to the Department of Physics of the University of Milano thanks to a fellowship supported by the Cariplo Foundation and the Landau Network—Centro Volta.

REFERENCES

1. R. C. Davidson, *An Introduction to the Physics of Nonneutral Plasmas* (Addison-Wesley, Redwood City, 1990).
2. R. B. Miller, *An Introduction to the Physics of Intense Charged Particle Beams* (Plenum, New York, 1982).
3. J. H. Malmberg and J. S. deGrassie, Phys. Rev. Lett. **35**, 577 (1975).
4. L. Brillouin, Phys. Rev. **67**, 260 (1945).
5. A. J. Theiss, R. A. Mahaffey and A. W. Trivelpiece, Phys. Rev. Lett. **35**, 1436 (1975).
6. L. D. Landau, E. M. Lifshitz, *A Course in Theoretical Physics. Vol. 2: Classical Theory of Fields*, (Pergamon Press, Oxford, 1971).
7. I. Kotelnikov, M. Romé and R. Pozzoli, Phys. Lett. A **372**, 1445 (2008).
8. I. Kotelnikov, M. Romé and R. Pozzoli, Phys. Lett. A **372**, 2450 (2008).

Stability of Non-Neutral Plasma Cylinder Consisting of Magnetized Cold Electrons and of Small Density Fraction of Ions Born at Rest: Non-Local Analysis

Y. N. Yeliseyev

Institute of Plasma Physics, National Science Center
"Kharkov Institute of Physics and Technology", 61108 Kharkov, Ukraine

Abstract. The non-local stability problem of the plasma cylinder, filled with "cold" magnetized rigidly rotating electrons, and a small density fraction of ions, is solved. The ions are supposed to be born at rest by ionization of background gas. The study is based on the kinetic description of ions. The equilibrium distribution function, taking into account the peculiarity of ions birth, is used. The radial electric field is caused by space charge of non-neutral plasma. The dispersion equation for plasma eigen frequencies is obtained analytically. It is valid within the total admissible range of values of electric and magnetic fields. Normalized eigen frequencies ω'/Ω_i are calculated for the basic azimuth mode $m=1$ ($\omega' = \omega - m\omega_+^+$, $\omega_+ = (-\omega_{ci} + \Omega_i)/2$, $\Omega_i = (\omega_{ci}^2 - 4eE_r/m_i r)^{1/2}$ is called the "modified" ion cyclotron (**MIC**) frequency), for the density fraction of ions of atomic nitrogen $f = N_i/n_e = 0.01$ and are presented in graphic form versus parameter $2\omega_{pe}^2/\omega_{ce}^2$. The spectra of oscillations ω'/Ω_i consist of the family of electron Trivelpiece – Gould (**TG**) modes and of the families of MIC modes. The frequencies of MIC modes are located in a small vicinity of harmonics of the MIC frequency Ω_i above and below the harmonic. The TG modes in non-neutral plasma fall in the region of MIC frequencies Ω_i and interact strongly with MIC modes. The slow TG modes become unstable near the crossings with non-negative harmonics of MIC frequencies. The instabilities have a resonant character. The lowest radial TG mode has a maximum growth rate at crossing with a zero harmonic of Ω_i (($\operatorname{Im}\omega'/\Omega_i)_{\max} \approx 0,074$). The growth rates of MIC modes are much lower (($\operatorname{Im}\omega'/\Omega_i)_{\max} \leq 0,002$). Their instability has a threshold character. The instabilities of TG and MIC modes take place mainly at the values of parameter $2\omega_{pe}^2/\omega_{ce}^2$, corresponding to strong radial electric fields ($\omega_{ci}^2 \ll |eE_r/m_i r|$), in which the ions are unmagnetized. The oscillations of small amplitude are seen on some frequency dependencies of MIC modes. They are similar to oscillations on dispersion curves of electron waves in metals and are caused by the similarity between the ion equilibrium distribution function and the degenerate Fermi - Dirac one. The results obtained give the solution to the stability problem discussed by R.H. Levy, J.D. Daugherty and O. Buneman [*Phys. Fl.* **12**, 2616 - 2629 (1969)] for a special case of plasma bounding directly with metal casing and possessing the volumetric eigen modes only.

Keywords: Non-Neutral Plasma, Crossed Fields, Unmagnetized Ions, Large Ion Larmor Radii, Ion Cyclotron Instability, Trivelpiece – Gould modes.
PACS: 52.20.Dq; 52.25.Dg; 52.27.Jt; 52.35.-g; 52.35.Fp; 52.35.Qz; 41.20.Cv.

INTRODUCTION

The non-local studying of the stability of plasma cylinder, placed in crossed longitudinal magnetic (B) and radial electrical ($E_r < 0$) fields, was carried out in [1,2]. Running helical potential waves

$$\tilde{\Phi} = \tilde{\Phi}_m(r)\exp[i(m\varphi + k_z z - \omega t)], \qquad (1)$$

having the frequencies ω - about a frequency of radial oscillations of ion in crossed fields Ω_i, were examined. The frequency $\Omega_i = \left(\omega_{ci}^2 - 4eE_r/m_i r\right)^{1/2}$ is also called "modified" ion cyclotron (MIC) frequency (m, k_z are azimuth number and longitudinal wave vector, ω_{ci} is the cyclotron frequency of ion).

Plasma consists of magnetized electrons, homogeneously distributed on radius ($n_e = const$, $r < a$) and rigidly rotating in crossed fields around an axis of the cylinder with frequency $\omega_e = -cE_r/(Br)$, and of a small addition of ions of one sort, which were born under ionization of residual gas by electron impact and move further collisionless. In the strong radial electric field ($\omega_{ci}^2 \ll |eE_r/m_i r|$, $E_r = -d\Psi(r)/dr$) such ions are unmagnetized. They move along large trajectories strongly elongated on radius. For adequate description of such ions the kinetic one is necessary. The peculiarity of ions is that at the initial moment after their birth they are at rest. This peculiarity determinates uniquely the equilibrium distribution function of ions. Its form was found in [3,4] and was used in [1,2]:

$$F(\varepsilon_\perp, M, v_z) = \frac{N_i}{T_i}\frac{m_i}{\omega_{ci}} Y\left(e\Psi(a) - \varepsilon_\perp\right) \cdot \delta\left(\varepsilon_\perp - \omega_{rot} M\right) \cdot \delta(v_z). \qquad (2)$$

Here N_i is a density of born ions ($N_i = const$, $r < a$), a distribution of an electrostatic potential $\Psi(r)$ is a square-law function of radius ($\Psi(r) = \Psi(a)(r^2/a^2)$, $\Psi(a) > 0$), $T_i = 2\pi/\Omega_i = const$ is the period of radial oscillations of an ion in crossed fields, Y is the Heaviside step function, δ is the Dirac delta-function, $\varepsilon_\perp$, M are transversal energy and generalized angular momentum of an ion, v_z is a longitudinal velocity. Factor $Y\left(e\Psi(a) - \varepsilon_\perp\right)$ in (2) reflects the fact of birth of ions only in a volume of the cylinder $r < a$. It makes the distribution function (2) similar to the degenerate Fermi - Dirac distribution function. Factor $\delta(\varepsilon_\perp - \omega_{rot} M)$ reflects the fact of ion birth at rest and makes the distribution functions (2) similar to «rigid rotator» one [5]. However, it does not belong to this type in a whole volume of plasma because of the additional dependence on energy $\varepsilon_\perp$ contained in factor $Y\left(e\Psi(a) - \varepsilon_\perp\right)$.

Factor ω_{rot} in (2) is not a free parameter. Its value is also uniquely determined by the fact of ion birth at rest and is equal to

$$\omega_{rot} = -cE_r/(Br) = const > 0. \qquad (3)$$

This value coincides formally with a rotation frequency of a charged particle in crossed fields, when radial electrical field is weak ($\omega_{ci}^2 \gg |eE_r/m_i r|$). However, expression (3) is valid in strong radial electrical field, as well.

The equilibrium distribution function of plasma electrons was chosen in [1, 2] as a Maxwellian one "with a shift" equal to $V_\varphi = \omega_e r$. It is the equilibrium of a «rigid rotator»-type [5]. Electrons were supposed to be "hot", $(\omega - m\omega_e)/(k_z v_{Te}) \approx m\omega_e/(k_z v_{Te}) \ll 1$ (v_{Te} is a thermal velocity of an electron). In the present paper the stability of plasma having "cold" electrons is studied,

$$(\omega - m\omega_e)/(k_z v_{Te}) \approx (m\omega_e)/(k_z v_{Te}) \gg 1. \qquad (4)$$

DISPERSION EQUATION OF PLASMA OSCILLATIONS

The non-local consideration of plasma stability is carried out in the same way as in [1, 2] - by the joint solving of the Vlasov - Poisson equations. Plasma is held within a metal casing (anode) of the same radius a, as the radius of plasma. Unknown radial function $\tilde{\Phi}_m(r)$ is represented in Bessel function expansion:

$$\tilde{\Phi}_m(r) = \sum_{l=1}^{\infty} C_m^l J_m(\kappa_{m,l} r/a)/N_m^l \qquad (5)$$

(factors C_m^l are the expansion coefficients, $N_m^l = (1/\sqrt{2})|J_{|m|+1}(\kappa_{m,l})|$ are the norms of Bessel functions, $\kappa_{m,l}$ - l-s root of Bessel function J_m, $J_m(\kappa_{m,l}) = 0$). Repeating the procedure stated in [1, 2], we obtain the homogeneous set of the linear equations for expansion coefficients C_m^l:

$$(L_{kl} - A_{kl})C_m^l = 0. \qquad (6)$$

The summation on subscript l is carried out in (6). A diagonal matrix L_{kl} equals

$$L_{kl} = (\kappa_{m,k}^2 \varepsilon_1^e + k_z^2 a^2 \varepsilon_3^e)\delta_{kl}, \qquad (7)$$

δ_{kl} is the Kronecer symbol. The components of dielectric permeability tensor of the rotating "cold" electron plasma $\varepsilon_{1,3}^e$ equal [5]

$$\varepsilon_1^e = 1 - \frac{\omega_{pe}^2}{(\omega - m\omega_e)^2 - \Omega_e^2}, \quad \varepsilon_3^e = 1 - \frac{\omega_{pe}^2}{(\omega - m\omega_e)^2}, \qquad (8)$$

ω_{pe} is the Lengmuir frequency of electrons, Ω_e is the "modified" electron cyclotron frequency in the crossed fields. In the present work (unlike [1, 2]) we assume, that the radial electric field is caused by a space charge of electrons and ions of plasma: $-(2e/m_e)E_r = \omega_{pe}^2(1-f)r$ ($f = N_i/n_e$ is a factor of charge neutralization). Setting the goal to carry out the studying of plasma stability within the whole admissible range of radial electric field changing, we use the general expressions for frequencies ω_e and Ω_e in (8), valid at arbitrary strengths of radial electric fields, [5]

$$\omega_e = (1/2)(|\omega_{ce}| - \Omega_e), \Omega_e = |\omega_{ce}|\left(1 - (2\omega_{pe}^2/\omega_{ce}^2)(1-f)\right)^{1/2}, 1 \geq (2\omega_{pe}^2/\omega_{ce}^2)(1-f) \geq 0. \qquad (9)$$

The last inequality in (9) determines the range of admissible values of parameter $2\omega_{pe}^2/\omega_{ce}^2$. It is an important parameter of the problem considered, which determines the relationship between electric and magnetic fields, and which determines whether ions are magnetized or not. Electrons are assumed to be magnetized and homogeneously distributed on radius in any considered fields.

The applicability condition of "cold" electrons approximation (4), expressed through parameter $2\omega_{pe}^2/\omega_{ce}^2$, has the form

$$2\omega_{pe}^2/\omega_{ce}^2 \gg (4/m)k_z\rho_{Le}/(1-f), \qquad (10)$$

where $\rho_{Le} = v_{Te}/\omega_{ce}$ is the electron Larmor radius. The condition of strong radial electric field for ions ($\omega_{ci}^2 \gg |eE_r/m_ir|$), looks as follows

$$2\omega_{pe}^2/\omega_{ce}^2 \gg 4(m_e/m)/(1-f). \qquad (11)$$

As follows from (11), ions cannot be treated as magnetized almost within the whole admissible range of changing of parameter $2\omega_{pe}^2/\omega_{ce}^2$.

Symmetric matrix

$$A_{kl} = \frac{\omega_{pi}^2}{N_m^l N_m^k} \sum_{p=-\infty}^{\infty} \left\{ \frac{1}{\omega_+^2} \frac{p}{\left(\frac{\omega'}{\Omega_i}-p\right)} J_{m+p}(z_l^-)J_{m+p}(z_k^-)J_p(z_l^+)J_p(z_k^+) + \right. \qquad (12)$$

$$+ \left[\frac{m}{\omega_-^2} + p\left(\frac{1}{\omega_-^2} - \frac{1}{\omega_+^2}\right)\right] \cdot \frac{1}{\left(\frac{\omega'}{\Omega_i}-p\right)} \int_0^1 dx \frac{d}{dx}\left[J_{m+p}(z_l^-x)J_{m+p}(z_k^-x)\right]J_p(z_l^+x)J_p(z_k^+x) +$$

$$\left. + \frac{1}{\Omega_i^2} \frac{k_z^2 a^2}{\left(\frac{\omega'}{\Omega_i}-p\right)^2} \int_0^1 x dx J_{m+p}(z_l^-x)J_{m+p}(z_k^-x)J_p(z_l^+x)J_p(z_k^+x) \right\}$$

describes the non-local contribution of ions and is the same as matrix A_{kl} in [1,2]. In (12) $\omega_{pi}^2 = 4\pi e^2 N_i/m_i$, $z_l^\pm = \kappa_{m,l}|\omega_\pm|/\Omega_i$, $\omega_\pm = (-\omega_{ci} \pm \Omega_i)/2$, $\omega' = \omega - m\omega_+$ is the frequency of a wave in a frame of reference, rotating with a "slow" rotation frequency of ions in crossed fields ω_+.

Frequencies of eigen plasma oscillations ω'/Ω_i are determined by the condition of consistency of the set of equations (6):

$$\det(L_{kl} - A_{kl}) = 0. \qquad (13)$$

Condition (13) is a dispersion equation of plasma eigen frequencies. Expressions (6) - (13) are valid within the whole admissible range of values of electric and magnetic fields, for both magnetized and unmagnetized ions, at arbitrary azimuth number m and longitudinal wave vector k_z, within the whole range of "modified" ion cyclotron frequencies $\omega' \sim \Omega_i$.

TRIVELPIECE - GOULD MODES

In contrast to plasma with "hot" electrons studied in [1, 2], plasma consisting of "cold" electrons has eigenmodes in the absence of ions. These are the modes of volumetric oscillations of Trivelpiece-Gould **(TG)** [6] in rotating electron plasma. Their frequencies are determined from the dispersion equation $\det(L_{kl}) = 0$ and are equal in the laboratory frame of reference [5]

$$\omega = m\omega_e \pm \left\{ (1/2)(\Omega_e^2 + \omega_{pe}^2) \pm \left[(1/4)(\Omega_e^2 + \omega_{pe}^2)^2 - \omega_{pe}^2 \Omega_e^2 \cos^2\theta \right]^{1/2} \right\}^{1/2}. \quad (14)$$

In (14) $\cos^2\theta = k_z^2/(k_z^2 + k_\perp^2)$, $k_\perp^2 = \kappa_{m,l}^2/a^2$. Expression (14) determines two families of radial TG modes: high-frequency ("TG+") and low-frequency ("TG-") modes (marks plus and a minus in braces). Inside each family expression (14) determines fast and slow modes (marks plus and a minus before braces).

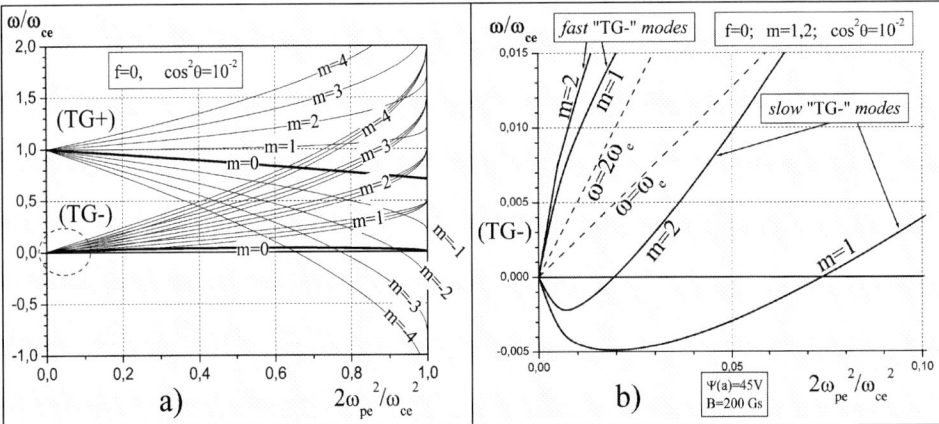

FIGURE 1. The Trivelpiece-Gould modes: a) within the whole admissible range of values of $2\omega_{pe}^2/\omega_{ce}^2$ parameter for azimuth wave numbers $|m| \leq 4$ and b) the same dependences in the region of low frequencies and small values of parameter $2\omega_{pe}^2/\omega_{ce}^2$ for azimuth wave numbers $|m| = 1, 2$. Formula (14) evaluation. In fig. 1b under the value of parameter $2\omega_{pe}^2/\omega_{ce}^2 = 0.05$ the corresponding values of a potential difference $\Psi(a)$ and of magnetic field strength B are shown as an example.

Usually eigen frequencies are represented versus longitudinal wave number [5] k_z. In the present paper all frequencies are represented versus parameter $2\omega_{pe}^2/\omega_{ce}^2$. This parameter (but not k_z) can change in experiment. Fig. 1 illustrates the behavior of solutions (14) versus this parameter. We'll dwell on behavior of modes of the "TG-" family. As it is seen from fig. 1a and in more details from a large-scale fig. 1b, at $2\omega_{pe}^2/\omega_{ce}^2 \to 0$ frequencies of all modes of this family aspire to zero and can interact in this area with ions. The frequency of the fast mode is positive. At increasing of the $2\omega_{pe}^2/\omega_{ce}^2$ parameter it quickly grows and leaves the area of ion frequencies. The frequency of a slow mode at $2\omega_{pe}^2/\omega_{ce}^2 \to 0$ is negative (fig. 1b). Its frequency sharply decreases at increasing of the $2\omega_{pe}^2/\omega_{ce}^2$ parameter, but quickly reaches the minimum. Then it grows, passes through zero frequency into the area of positive frequencies and leaves the area of ion frequencies. It is interesting to note that TG modes of neutral plasma, the electrons of which drift along a magnetic field, have the same peculiarity of frequency behavior, but versus a longitudinal wave vector k_z [6].

The value of parameter $2\omega_{pe}^2/\omega_{ce}^2$, at which the zero frequency of a slow mode is reached, at $\cos^2\theta \ll 1$ equals

$$2\omega_{pe}^2/\omega_{ce}^2 = \left(2\omega_{pe}^2/\omega_{ce}^2\right)_0 \approx (8/m^2)\cos^2\theta(1-f)^{-2}. \qquad (15)$$

Inside the interval $0 < 2\omega_{pe}^2/\omega_{ce}^2 \lesssim (2\omega_{pe}^2/\omega_{ce}^2)_0$ the frequency of slow TG mode of rotating electron plasma (14) remains within a range of ion frequencies $\omega \sim \Omega_i$ and this mode can effectively interact with ions. A small (at $\cos^2\theta \gg m_e/m_i$) value of parameter $(2\omega_{pe}^2/\omega_{ce}^2)_0$ (15) corresponds to a strong radial electric field for ions (11).

SPECTRA OF PLASMA OSCILLATIONS

The solutions of the dispersion equation (13), when a small addition of ions is present in plasma, were determined numerically. Matrix elements in equation (13) and the normalized frequencies ω'/Ω_i depend on five parameters, accordingly: $2\omega_{pe}^2/\omega_{ce}^2$, m_e/m_i, $f = N_i/n_e$, $k_z a$, m. Calculations were carried out for the basic azimuth mode $m = +1$, the ion mass is chosen equal to ion mass of atomic nitrogen ($m_i = 14$ a.u), $k_z a = 0,1$, $f = N_i/n_e = 0,01$. The normalized spectra of plasma oscillations ω'/Ω_i are submitted versus parameter $2\omega_{pe}^2/\omega_{ce}^2$, which was set in the range $10^{-6} \leq 2\omega_{pe}^2/\omega_{ce}^2 \leq 1/(1-f)$. Solutions of the dispersion equation (13) were determined by the same method, as in [1, 2]. In matrix element A_{kl} (13) the summands were taken into account in the sum over p, having $-80 \leq p \leq 80$. In infinite set of the equations (6) 30 equations were kept ($1 \leq k \leq 30$), in the sum on subscript l, 30 summands ($1 \leq l \leq 30$) were taken into account. Zeros of the determinant of a 30×30 matrix were determined. Replacement of an infinite matrix by a finite one corresponds to the Galerkin projective method for solving problems.

Results of calculations are submitted in fig. 2-6. In fig. 2 the general picture of behavior of normalized plasma eigen frequencies ω'/Ω_i is submitted. In the range of ion frequencies the spectrum consists of the family of low-frequency "GT-" modes (further designated simply as GT) and of the families of MIC modes. The TG modes are well visible. Their frequencies are close to the analytical solution (14). At the chosen values of parameters (m, $k_z a$, f, an ion mass) the TG modes cross harmonics of MIC frequency Ω_i, having numbers $\omega'/\Omega_i = n \geq -2$. The MIC modes, because of small density of ions, are located in a small vicinity of harmonics of Ω_i (positive, negative and zero) and are not seen in fig.2. They are traced in details in fig. 3-6.

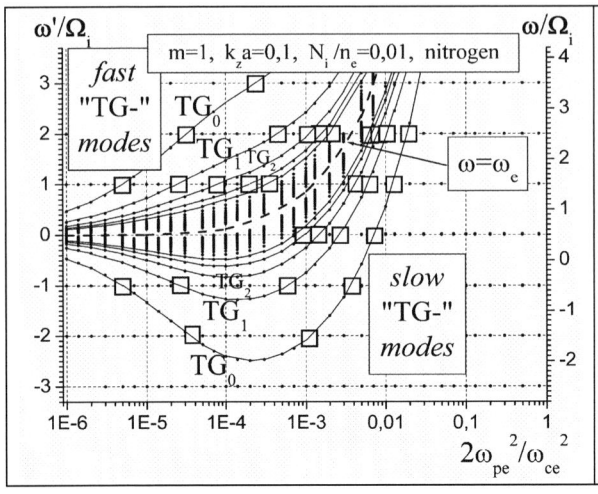

FIGURE 2. The frequencies of "TG-" modes versus parameter $2\omega_{pe}^2/\omega_{ce}^2$ in a frame of reference rotating with slow frequency of ion rotation ω_-. The formula (13) evaluation. Fast TG modes lay above dashed line $\omega = \omega_e$, slow – below it. The subscripts denote the numbers of radial modes. Square boxes designate crossings TG modes with MIC modes. The true behavior of modes in a vicinity of harmonics of Ω_i is submitted in details in fig. 3-6.

The behavior of modes in a vicinity of the $\omega'/\Omega_i = n = -2$ harmonic of MIC frequency is submitted in details in fig. 3 a - d. Only the zero radial TG_0 mode crosses this harmonic (fig. 2). Below the harmonic the GT_0 mode turns into the MIC_0 mode and both modes remain stable (fig. 3a). Above the harmonic, at crossing the TG_0 and MIC_0 modes, instability arises. It exists inside the $2\omega_{pe}^2/\omega_{ce}^2$ parameter domain, laying between crossings of GT_0 mode with this harmonic. Dependence of a real part of frequency of unstable oscillations versus parameter $2\omega_{pe}^2/\omega_{ce}^2$ has the same character as the stable MIC mode, therefore we speak about the instability of a MIC_0 mode. Its growth rate (fig.3b) is small, $\text{Im}(\omega'/\Omega_i) < 10^{-4}$.

In fig. 3a near the value of parameter $2\omega_{pe}^2/\omega_{ce}^2 \approx 0.894$, the peculiarity of eigen frequency behavior is seen: all radial MIC modes aspire to value $\omega'/\Omega_i = -2$. To the left of the point of peculiarity all eigen frequencies are real, to the right-complex, having the real and imaginary parts (fig.3c,d). To the right, at a small distance from the peculiarity, frequencies of two modes, one of which is a zero radial MIC_0 mode, become real. The MIC_1 mode has a maximum growth rate, though it is small, $\text{Im}(\omega'/\Omega_i) < 4 \cdot 10^{-4}$. The peculiarity is caused by the fact that in point $2\omega_{pe}^2/\omega_{ce}^2 \approx 0.894$ the equality $m^2\omega_e^2 - \Omega_e^2 \approx 0$ is fulfilled and the component of tensor of dielectric permeability ε_1^e (8) has a pole. To the left of a pole $\varepsilon_1^e > 0$, to the right-$\varepsilon_1^e < 0$. Matrix elements L_{kl} (7) change a sign in this point. This is why the instability is considered to be a reactive type [7]. As it is seen from fig. 3c,d, to the right of a pole the imaginary part of frequency $\text{Im}(\omega'/\Omega)$ arises faster than real. The instability must be caused by the third term in the matrix element A_{kl} (12), proportional to $k_z^2 a^2$.

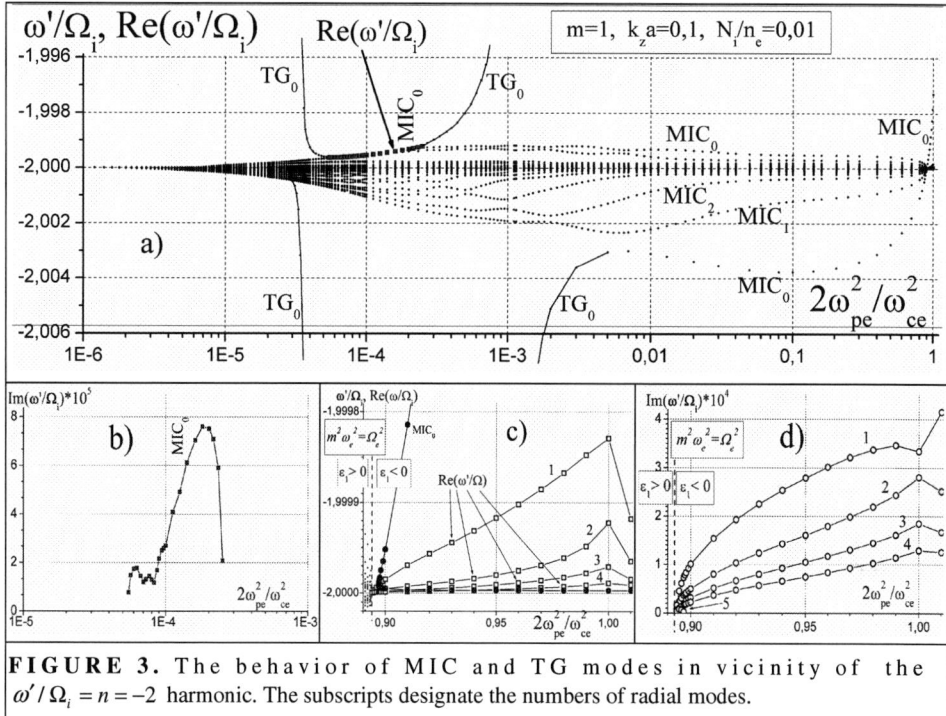

FIGURE 3. The behavior of MIC and TG modes in vicinity of the $\omega'/\Omega_i = n = -2$ harmonic. The subscripts designate the numbers of radial modes.

In fig. 3a the oscillations of small amplitude are appreciable on frequency dependences of MIC modes. Such oscillations are appreciable also on dependences of MIC modes in the vicinity of the $\omega'/\Omega_i = n = -1$ harmonic (fig. 4a) at greater magnification. We suppose they are similar to oscillations on dispersion curves of electron waves in metals [8] and are caused by the similarity between the ion distribution function (2) and the degenerate Fermi - Dirac one. The amplitude of oscillations is small because of the small density of ions.

The $\omega'/\Omega_i = n = -1$ harmonic of MIC frequency is crossed by TG$_0$ and TG$_1$ radial modes (fig. 2 and 4 a-d). The behavior of MIC modes in a vicinity of this harmonic is similar to their behavior near the $n = -2$ harmonic. But there are also differences. As it is seen from fig. 4a, frequencies of MIC modes are more distant from the $n = -1$ harmonic ($|\omega'/\Omega_i + 1| \leq 0,014$), than in the case of the $n = -2$ harmonic ($|\omega'/\Omega_i + 2| \leq 0,004$). There is a sharp increase of frequency at the right edge of the instability region. The maximal growth rates of the MIC$_0$ and MIC$_1$ modes are located near these edges (fig. 4b) and reach values $\mathrm{Im}(\omega'/\Omega_i) \approx 1,9 \cdot 10^{-3}$ and $\mathrm{Im}(\omega'/\Omega_i) \approx 1,4 \cdot 10^{-3}$ respectively.

In fig. 4a the peculiarity of eigen frequency behavior, caused by the pole ε_1^e (8), is present near the value of parameter $2\omega_{pe}^2/\omega_{ce}^2 \approx 0.897$. The behavior of frequencies and growth rates in the vicinity of peculiarity (fig. 4c, d) is similar to their behavior near the $\omega'/\Omega_i = -2$ harmonic. The maximal growth rate in fig. 4d ($\text{Im}(\omega'/\Omega_i) \approx 1 \cdot 10^{-3}$) is more than the growth rate of mode MIC$_1$ in fig. 3d.

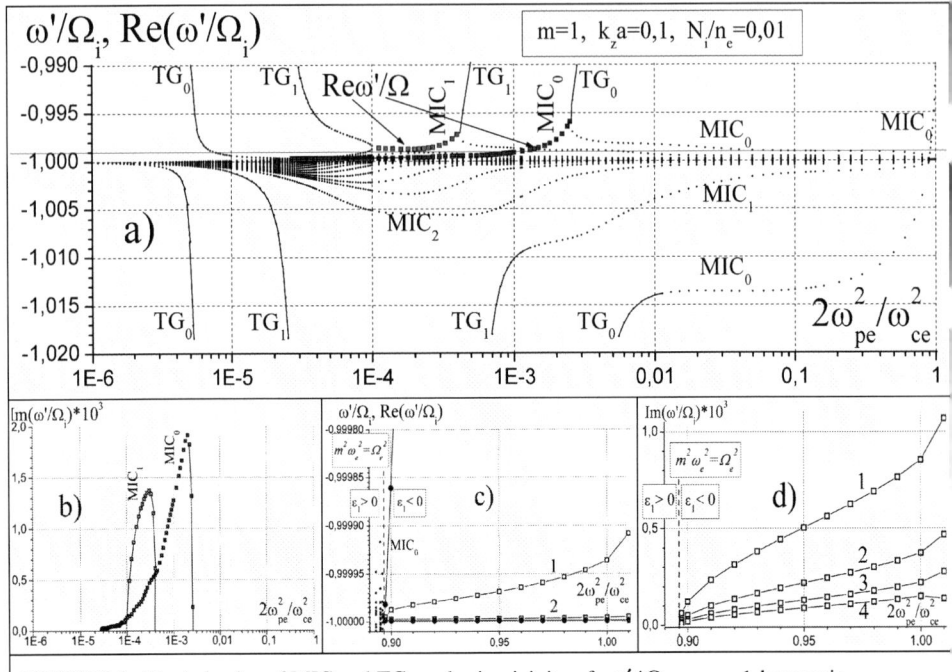

FIGURE 4. The behavior of MIC and TG modes in vicinity of $\omega'/\Omega_i = n = -1$ harmonic.

The $\omega'/\Omega_i = n = 0$ harmonic of MIC frequency is crossed by all slow low-frequency radial modes TG (fig. 2). However, the Galerkin method is inapplicable for the highest radial TG and MIC modes. In fig. 5 a-d the results of calculation are submitted in a region $4 \cdot 10^{-4} < 2\omega_{pe}^2/\omega_{ce}^2 < 1$, where the Galerkin method is confidently applicable. The behavior of modes near to zero harmonic of Ω_i differs from behavior of modes near to negative ones. In a vicinity of a zero harmonic both GT modes, and MIC modes are unstable (fig. 5a, b). The instability of GT modes takes place in a wide range of frequency near to zero harmonic ($|\omega'/\Omega_i| \leq 0,08$) with growth rate, by more than the order of magnitude exceeding the growth rate of MIC modes. The growth rate of TG$_0$ mode is the largest ($\text{Im}(\omega'/\Omega_i) \leq 0,074$). The instabilities of MIC modes take place below the harmonic ($\omega'/\Omega_i < 0$) and to the left of a point of crossing of corresponding TG mode with the harmonic. The maximal growth rate of MIC$_0$ mode equals $\text{Im}(\omega'/\Omega_i) \approx 2 \cdot 10^{-3}$.

A peculiarity of eigen frequency behavior, caused by the pole ε_1^e (8), is near the value of parameter $2\omega_{pe}^2/\omega_{ce}^2 \approx 0.9$ in fig. 5a. The behavior of frequencies and of growth rates to the right of the peculiarity is shown in fig. 5c, d. The MIC$_1$ mode has a maximal growth rate (fig. 5d), $\text{Im}(\omega'/\Omega_i) \approx 1 \cdot 10^{-3}$.

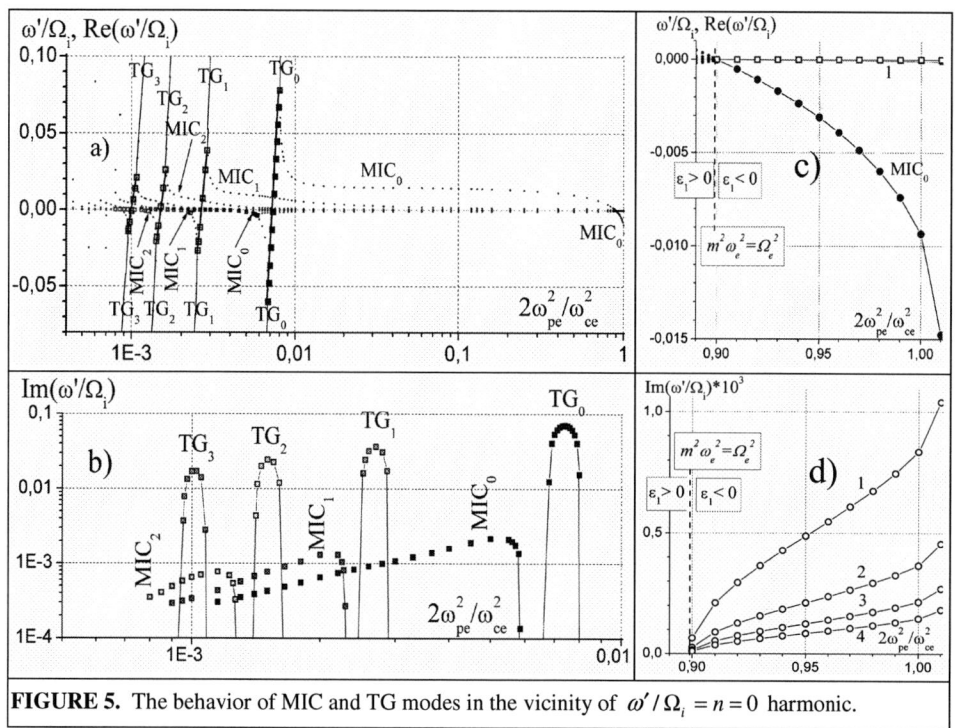

FIGURE 5. The behavior of MIC and TG modes in the vicinity of $\omega'/\Omega_i = n = 0$ harmonic.

The $\omega'/\Omega_i = n = +1$ harmonic and other positive harmonics are crossed by all low-frequency radial TG modes both fast, and slow (fig. 2, 6). The $2\omega_{pe}^2/\omega_{ce}^2$ parameter domain, where Galerkin method is inapplicable, is shaded in fig. 6a. As well as in vicinities of the $\omega'/\Omega_i = n = 0$ harmonic, slow TG modes are unstable (fig. 6a). Their growth rates (fig. 6b) are less ($\text{Im}(\omega'/\Omega_i) \leq 0,042$) than they are in the vicinity of a zero harmonic. The fast TG modes are stable. The MIC modes are unstable below the $\omega'/\Omega_i = n = +1$ harmonic and to the left of the point of crossing of corresponding slow TG mode and to the right of the point of crossing of fast TG mode. The growth rate of the MIC_0 mode is small, $\text{Im}(\omega'/\Omega_i) \leq 10^{-4}$. The growth rates of higher radial modes (MIC_1, MIC_2,...) are by the order of magnitude larger, $\text{Im}(\omega'/\Omega_i) \approx 10^{-3}$.

A peculiarity of eigen frequency behavior, caused by the pole ε_1^e (8), is near to value of parameter $2\omega_{pe}^2/\omega_{ce}^2 \approx 0.904$ in fig. 6a. The frequencies and growth rates to the right of peculiarity are submitted in fig. 6 c, d. Their behavior is similar to behavior of frequencies and growth rates in the vicinity of zero harmonic Ω_i (fig. 5c, d). The growth rate of MIC_1 mode is a maximal one but it is small (fig. 6d), $\text{Im}(\omega'/\Omega_i) \leq 3 \cdot 10^{-4}$.

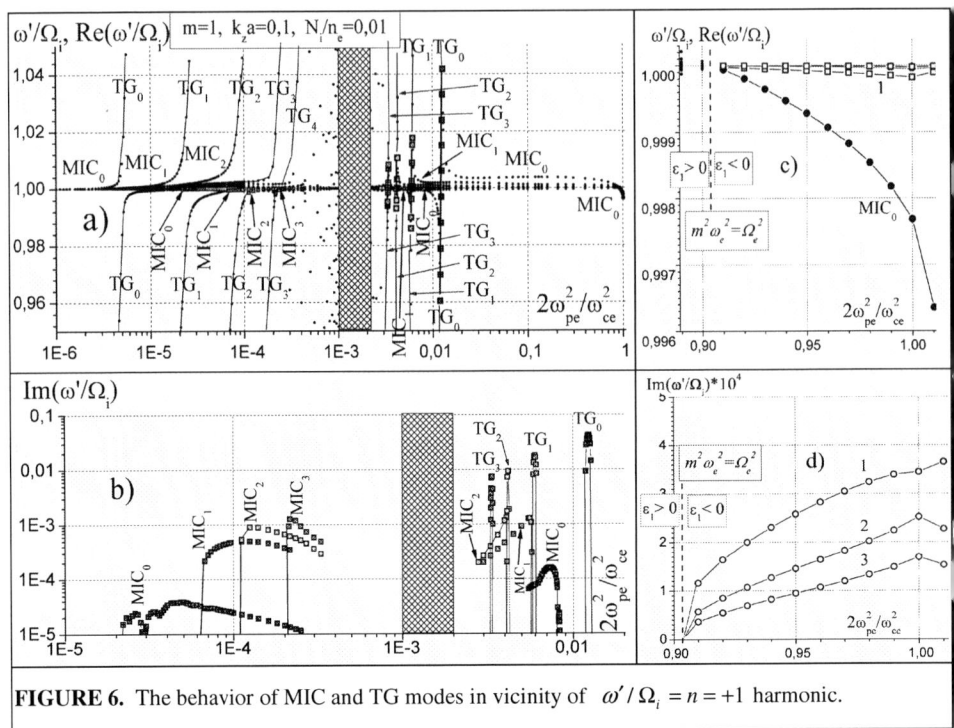

FIGURE 6. The behavior of MIC and TG modes in vicinity of $\omega'/\Omega_i = n = +1$ harmonic.

CONCLUSIONS

The non-local stability problem of non-neutral plasma cylinder, consisting of magnetized "cold" electrons and of a small additive of ions, born under ionization of atoms, molecules of residual gas by electron impact, is solved. The research carried out is based on the kinetic description of ions, adequately taking into account the peculiarity of their birth.

The problem considered is a special case of a more general statement of problem, when there is a vacuum clearance between the plasma cylinder and a metal casing. In this case both surface and volumetric TG modes are present in a spectrum. In [9] the excitation of surface modes by magnetized ions was considered. Their frequencies under certain conditions fall in region of ion frequencies. In plasma completely filling a waveguide, surface modes are absent. Such plasma possesses only volumetric modes of oscillations. In non-neutral plasma because of Doppler shift, caused by rotation of electrons, low-frequency volumetric electron TG modes also fall in region of ion frequencies, where they can interact with MIC modes.

The dispersion equation of plasma oscillations (13) was obtained analytically. It is valid within the whole admissible range of change of electric and magnetic fields. The solutions of the dispersion equation (13) are found numerically and submitted versus the $2\omega_{pe}^2/\omega_{ce}^2$ parameter, describing the relationship between electric and magnetic fields and determining whether ions are magnetized or not.

The calculations showed, that the spectra of plasma oscillations consist of the family of volumetric electron low-frequency TG modes (fig. 2), having frequencies close to the analytical expression (14) for the frequencies of TG modes in pure electron plasma, and of the families of MIC modes. The latter have frequencies close to harmonics of MIC frequencies Ω_i, including zero harmonic, and are arranged above and below the harmonic, as in case of "hot" electrons [1,2].

The slow low-frequency TG modes are unstable in a vicinity of crossing with nonnegative harmonics of the MIC frequency Ω_i. Instability has resonant character. The TG$_0$ mode has the largest growth rate in a vicinity of crossing with a zero harmonic of the MIC frequency Ω_i. At the chosen numerical values of parameters it reaches the value $\text{Im}(\omega'/\Omega_i) \approx 0,074$. Fast low-frequency TG modes are stable. MIC modes are unstable inside the $2\omega_{pe}^2/\omega_{ce}^2$ parameter domain, laying between the crossings of harmonic with corresponding TG modes. Instability has threshold character. The growth rates of MIC modes ($\text{Im}(\omega'/\Omega_i) < 2 \cdot 10^{-3}$) are much less than the growth rates of TG modes.

At values of parameter $2\omega_{pe}^2/\omega_{ce}^2 \approx 0,9$, close to Brillouin limit, there is a peculiarity of frequency behavior of MIC modes, connected with the pole of component of tensor of dielectric permeability ε_1^e (8). To the right of this pole the reactive instabilities of MIC modes arise with maximal growth rate $\text{Im}(\omega'/\Omega_i) \lesssim 1 \cdot 10^{-3}$.

On some frequency dependences of MIC modes the oscillations of small amplitude are present. They are similar to oscillations on dispersion curves of electron waves in metals [8] and are caused by the similarity between the ion distribution function (2) and the degenerate Fermi - Dirac one.

The obtained spectra of plasma eigen oscillations shown in fig. 2-6 give the solution of a problem posed. The found instabilities of TG and MIC modes arise mainly in the region of a strong radial electric field, where ions are unmagnetized and non-local studying of plasma stability is necessary. Such studying is given in the present paper.

REFERENCES

1. Y. N. Yeliseyev, *Plasma Physics Reports* **32**, 927 - 936 (2006).
2. Y. N. Yeliseyev, "Equilibrium and Stability of Non-Neutral Plasma with Unmagnetized Ions Born at Rest and Moving along Large Orbits", in *Non-Neutral Plasma Physics IV*, edited by M. Drewsen et al., AIP Conference Proceedings 862, American Institute of Physics, Melville, NY, 2006, pp.108-115
3. Y. N. Yeliseyev, Y. A. Kirochkin and K. N. Stepanov, "Non-Local Theory of Ion Cyclotron Instability of a Plasma in Crossed Axial Magnetic and Strong Radial Electric Fields", in *1987 Intern. Conf. on Plasma Phys.*, ed. by A.G. Sitenko, Proc. Contrib. Papers, v.1, Kiev, Naukova dumka, 1987, pp. 26 – 29.
4. V. G. Dem'yanov, Y. N. Yeliseyev, Y. A. Kirochkin, A. A. Luchaninov, V. I. Panchenko and K. N. Stepanov, *Plasma Physics Reports* **14**, 494 - 500 (1988).
5. R. C. Davidson, *Theory of Nonneutral Plasma*, London: W.A. Benjamin inc., 1974.
6. A. W. Trivelpiece, R. W. Gould, *Journal of Applied Physics* **30**, 1784 - 1793 (1959).
7. G. Bekefi, *Radiation Processes in Plasmas*, New York: John Wiley and Sons, Inc., 1966.
8. D. G. Lominadze, *Cyclotron Waves in Plasma*, Tbilisi, "Metsniereba", 1975 (in Russian).
9. R. H. Levy, J. D. Daugherty and O. Buneman, *Phys. Fluids* **12**, 2616 - 2629 (1969).

SECTION IV

APPLICATIONS AND SPECIAL TOPICS

Radial compression of antiproton cloud for production of ultraslow antiproton beams

N. Kuroda*, Y. Nagata*, H.A. Torii*, D. Barna[†], J. Eades**, D. Horváth[†], M. Hori[‡], H. Imao[§], K. Komaki*, A. Mohri[§], M. Shibata[§] and Y. Yamazaki[§,*]

Institute of Physics, University of Tokyo, 3-8-1 Komaba, Meguro-ku, Tokyo, 153-8902, Japan
[†]*KFKI Research Institute of Particle and Nuclear Physics, H-1525 Budapest, Hungary*
**Department of Physics, University of Tokyo, 7-3-1 Hongo, Bunkyo-ku, Tokyo, 113-0033, Japan*
[‡]*Max-Planck-Institute für Quantenoptik, D-85748 Garching, Germany*
[§]*RIKEN, 2-1 Hirosawa, Wako-shi, Saitama, 351-0198, Japan*

Abstract. We report here the radial compression of a large number of antiprotons under ultrahigh vacuum conditions by applying a rotating electric field. The radial compression is a key technique for production of ultraslow antiproton beam extracting from an electromagnetic traps. Such beam will be applicable to synthesizing antiprotonic atoms and antihydrogen atoms.

Keywords: antiproton, non-neutral plasma, multiring trap
PACS: 39.10.+j, 52.27.Jt, 52.35.Fp

INTRODUCTION

Cooling and manipulation of a large number of antiprotons held in an electromagnetic trap are key techniques for synthesizing antihydrogen atoms and antiprotonic atoms [1, 2, 3]. We, MUSASHI sub-group in ASACUSA collaboration, achieved accumulation and cooling of antiprotons in an electromagnetic trap with at least 50 times higher efficiency than conventional method. Extracting these antiprotons from the trap and transporting them efficiently in the form of a beam is the next step not only towards synthesizing antihydrogen atoms for use in CPT symmetry test, but also for studying atomic collision dynamics. Since charged particles tent to follow magnetic field lines, a cloud of antiprotons should have a small radius in the trap for better focusing of extracted beams. We summarize our success of antiproton compression experiment [4] and also shortly describe the latest results.

EXPERIMENT

Antiprotons at 5 MeV from the CERN Antiproton Decelerator (AD) were slowed down to 115 keV by Radio Frequency Quadrupole Decelerator (RFQD) and then were accumulated and cooled in a so-called multiring trap (MRT) [5]. Figure 1 shows a schematic view of the experimental setup downstream of the RFQD: the MRT housed in a 2.5 T superconducting solenoid, the transport beam line, and a microchannel plate with a delay line anode used as a position sensitive detector (PSD). The MRT consists of 14 ring electrodes, the central five of them being used to form a harmonic potential along the trap axis [6] kept an ultrahigh vacuum condition ($\leq 10^{-10}$ Pa) to avoid loss of antiprotons via anniliation with residual gasses. One of the ring electrodes was azimuthally segmented into four parts to which a radio frequency (rf) field with phases shifted by $\pi/2$ for generating a rotating electric field. The lower left part of Fig. 1 shows the potential distribution along the axis. The potential of the MRT V_{MRT} corresponded to the kinetic energy of the extracted antiproton beam. The annihilation position along the beam line was determined by track detector consisting of two 2-m long plastic scintillator bars installed parallel to the beam line. The profile of antiproton beam was monitored by the PSD placed 1.75 m downstream from the center of the MRT. An additional plastic scintillator plate near the PSD identified the antiprotons by detecting passage of annihilation products such as energetic pions.

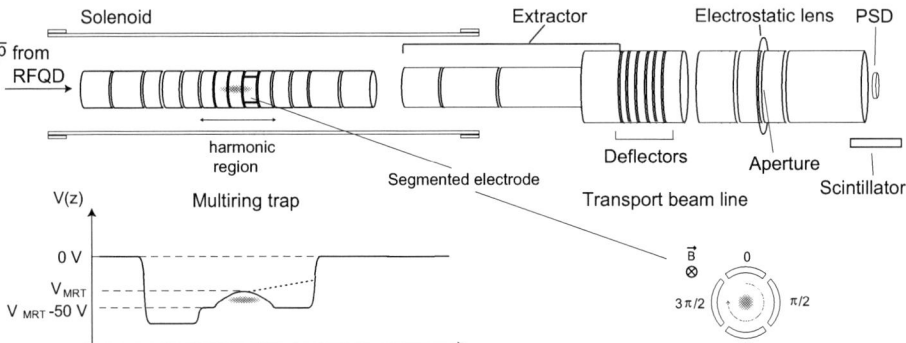

FIGURE 1. A schematic view of the main part of our experimental setup (not in scale) including the MRT with a cross section of the segmented electrode (right lower figure) and the transport beam line. Antiprotons ($\bar{p}$s) come from the left side. The left lower figure shows the potential distribution during the antiproton extraction.

After the cooling of antiprotons by collisions between preloaded electrons in the MRT, electrons were expelled from the trapping region by opening the trap potential as quick as to retain the antiprotons in the trap, but to allow the electrons to escape, these having much faster velocity than the electrons. We observed the number of electrons became almost 1/10 every one operation of the release. During the following experiments this release cycle was repeated for 6 times to guarantee the number of remained electrons as small as possible.

RESULTS AND DISCUSSIONS

Figure 2 (a)–(c) show the annihilation distribution along the transport beam line. When no rotating field was applied, most of the antiprotons annihilated around the extractor at the exit of the superconducting solenoid. Figures 2 (b) and (c) corresponds to the cases where a rotating electric field with its frequency 247 kHz was applied for 60 and 120 s. The annihilation peak around the PSD increased in size as t_r became longer. The corresponding 2D images on the PSD are shown in Figs.2 (d)–(f). Without the rotating field [case (d)], the image is dim with low intensity and show a depopulated region (hollow) in its center. This hollow gradually fills up as the field is applied. The antiproton cloud was therefore effectively compressed by the rotating electric field.

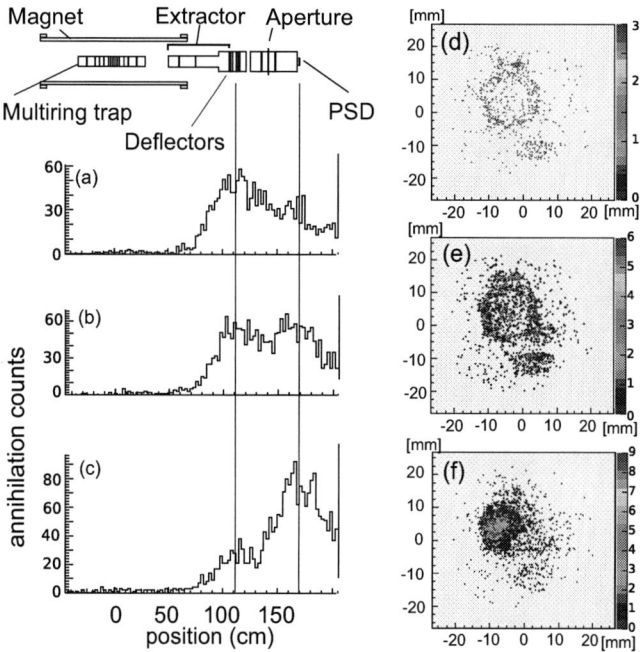

FIGURE 2. (a)–(c) The annihilation position of extracted antiprotons along the transport beam line for the compression time $t_r = 0$, 60, and 120 s. (d)–(f) The PSD images of extracted antiprotons for $t_r = 0$, 60, and 120 s. The frequency was $f = 247$ kHz with its peak-to-peak amplitude $V_r = 0.56$ V. The electron plasma used had a radius of 3.4 mm and was removed from the MRT before application of the rotating field and the extraction.

We performed trajectory simulations to relate the radial distribution of the antiproton cloud in the MRT with annihilation distribution along the beam line and the image on the PSD. Figures 3 (a) and (d) show the simulated position distribution of antiproton annihilation and the beam profile at the PSD, respectively, for an antiproton cloud with $a_{\bar{p}} = 3.4$ mm in the MRT, where extracted cloud has a profile expressed by a Gaussian function $\rho(r_{\bar{p}}) \propto \exp\{-(r_{\bar{p}}/a_{\bar{p}})^2/2\}$. The annihilation distribution in Fig. 2 (a) as well as the hollowed distribution of Fig. 2 (d) was reproduced reasonably by the simulation. When we define the transport efficeincy ε as the ration of the number of trapped antiprotons detected by the PSD to the number of trapped antiprotons, the calculated efficiency of $\varepsilon_{\text{sim}} = 0.06$ was also consistent with the experimental value of $\varepsilon_{\text{exp}} = 0.08$. Here $a_{\bar{p}} = 3.4$ mm corresponded to the size of the electron plasma used to cool antiprotons.

The PSD image in Fig. 2 (c) was almost reproduced with a profile in Fig. 3(g) for $a_{\bar{p}} = 0.25$ mm, as shown in Fig. 3(e). On the other hand, the shoulder around the annihilation position pf 100 cm in Fig. 2 (c) was not seen in Fig. 3 (b), and the simulated transport efficiency predicted $\varepsilon_{sim} \sim 1$, while the experimental value was 0.45 at most. We therefore conclude that the antiproton cloud in the MRT after the compression consists of two components. Figures 3(c), (f) with initial profile Fig. 3(h) superimposed of two components, well compressed 45% of $a_{\bar{p}} = 0.25$ mm and expanded 55% of $a_{\bar{p}} = 4.0$ mm, show the results of our simulation reproduces the observation in Figs. 2(c) and (f), respectively.

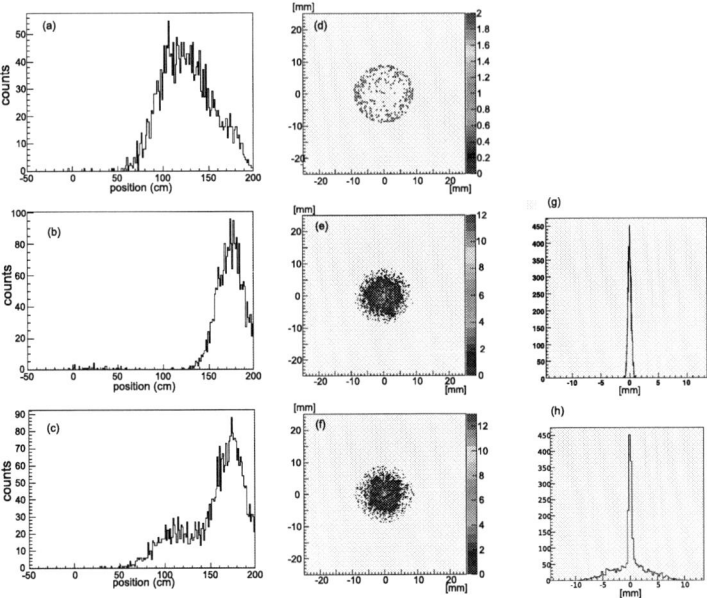

FIGURE 3. Simulated antiproton annihilation position distribution along the transport beam line [(a)–(c)] and antiproton beam profile at the PSD [(d)–(f)]. (a) and (d) $a_{\bar{p}} = 3.4$ mm, (b) and (e) $a_{\bar{p}} = 0.25$ mm as shown in (g), (c) and (f) a superposition of two Gaussian distributions, 45% of $a_{\bar{p}} = 0.25$ mm and 55% of $a_{\bar{p}} = 4.0$ mm as shown in (h).

We also survey parameters for compression like the frequency, amplitude, and compression time. Figure 4(a) shows the transport efficiency as a function of the rf frequency. The negative frequency refers to the counter rotating field. The transport efficiency appears to be roughly constant at around 0.3 for $f \geq 200$kHz. For $f < 100$ kHz and lower, the efficiency was approximately the same as it was without rotation. The amplitude dependence shown in Fig. 4(b) was also studied. This revealed that the transport efficiency was constant for $V_r \geq 50$ mV. The time dependency was shown in Fig. 4(c). For the case of fixed frequency, $f = 247$ kHz, the extraction efficiency monotonically increased from 0.08 for $t_r = 0$ to 0.4 – 0.45 for $t_r \geq 200$ s. The maximum number of transported antiproton was about 5×10^5 when 1.1×10^6 antiprotons were captured in the MRT. Some antiprotons were lost during the compression. The loss rate was twice as large with the rotating field and caused 10% loss of the trapped antiprotons for $t_r = 200$ s. When the frequency was swept from 100 kHz to 250 kHz, we found the antiproton cloud compression was preformed earlier as shown in open circles in Fig. 4(c).

These parameter dependencies shown in Fig. 4 indicated that the mechanism of the antiproton cloud observed here is phenomenologically similar to the strong drive compression observed for electron plasmas [6, 7, 8] rather than the observed sideband cooling for the proton cloud [9]. However the strong drive comressions were reported for olny light particles, electron and positron. In the case of antiproton clouds, where UHV conditions are essential to avoid loss by annihilation, and synchrotron radiaion cooling is completely negligible, it has been assumed that use of strong drive technique is difficult. The broad halo after the compression may become any evidence of antiproton compression without any coolant, but at the same time the confirmation of electron number after the kickout procedure is necessary.

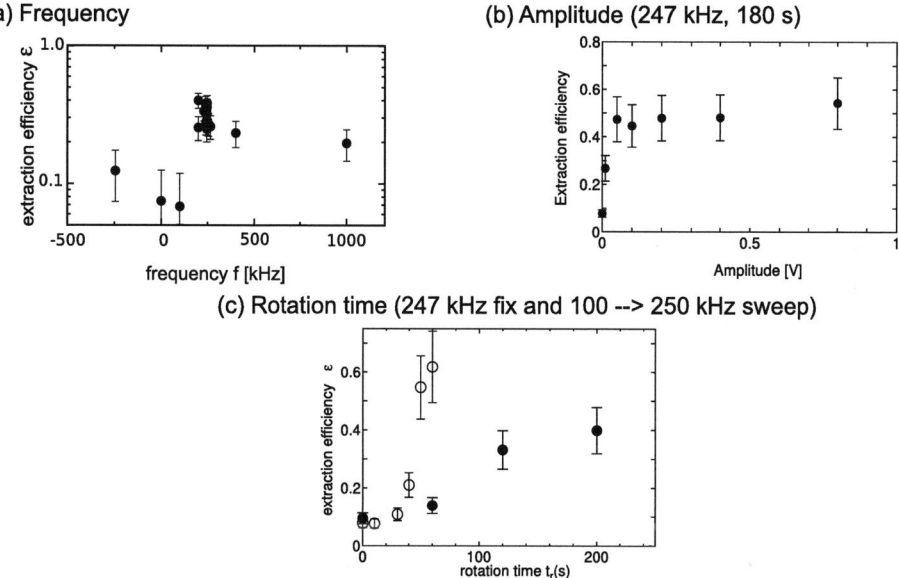

FIGURE 4. (a)The transport efficiency ε of antiprotons as a function of the rf frequency for t_r 120 s and $V_r = 0.56$ V. The error bars mainly come from the fluctuation of the antiproton beam intensity from the AD. (b) ε plotted against the amplitudes V_r for $f = 247$ kHz and $t_r = 180$ s. (c) ε against the compression time t_r for $f = 247$ kHz and swept rf from $f = 100$ kHz to 250 kHz (open circles).

We also demonstrated the compression with electrons just after we learned from the ALPHA collaboration's results [10]. In our results, antiprotons were observed to compress with electrons under a weak rotating field tuned the frequency for electron plasma compression.

CONCLUSION

In summary, we have found an effective compression method for antiprotons trapped in an ultrahigh vacuum of $p \leq 10^{-10}$ Pa. A rotating electric field over a broad frequency range successfully compressed a half of the antiproton cloud. The sweep of rf field helped to reduce compression time. These compression seemed to be similar to the strong drive regime, but elucidation of the compression mechanism in the UHV condition remains a subject for future studies.

Such produced ultraslow antiproton beams is now applicable to atomic collision experiments and started to make results [11].

ACKNOWLEDGMENTS

This work was supported by the Grant-in-Aid for Creative Scientific Research (10P0101) of the Japanese Ministry of Education, Culture, Sports, Science and Technology, Special Research Projects for Basic Science of RIKEN, and the Hungarian National Science Foundation (OTKA T033079).

REFERENCES

1. G. Gabrielse, *Adv. At. Mol. Opt. Phys.*, **50**, 155 (2004).
2. M. Charlton, J. Eades, D. Horváth, R. J. Hughes, and C. Zimmerman, *Phys. Rep.*, **241**, 65 (1994).
3. N. Kuroda, H.A. Torii, K. Yoshiki Franzen, Z. Wang, S. Yoneda, M. Inoue, M. Hori, B. Juhász, D. Horváth, H. Higaki, A. Mohri, J. Eades, K. Komaki, and Y. Yamazaki, *Phys. Rev. Lett.*, **94**, 023401 (2005).
4. N. Kuroda, H. Torii, M. Shibata, Y. Nagata, D. Barna, M. Hori, D. Horváth, A. Mohri, J. Eades, K. Komaki, and Y. Yamazaki, *Phys. Rev. Lett.*, **100**, 203402 (2008).
5. A. Mohri, H. Higaki, H. Tanaka, Y. Yamazawa, M. Aoyagi, T. Yuyama, and T. Michishita, *Jpn. J. Appl. Phys.*, **37**, 664 (1998).
6. N. Kuroda, H. Torii, M. Shibata, Y. Nagata, D. Barna, M. Hori, J. Eades, A. Mohri, K. Komaki, and Y. Yamazaki, "Control of plasmas for production of ultraslow antiproton beams," in *Workshop on Physics with Ultra Slow Antiproton Beams*, edited by Y. Yamazaki, and M. Wada, RIKEN, AIP, New York, 2005, vol. 793 of *AIP Conf. Proc.*
7. J. Danielson, and C. Surko, *Phys. Rev. Lett.*, **94**, 035001 (2005).
8. L.V. Jørgensen, M. Amoretti, G. Bonomi, P. Bowe, C. Canali, C. Cesar, M. Charlton, M. Doser, A. Fontana, M. Fujiwara, P. Genova, J. Hangst, R. Hayano, A. Kellerbauer, V. Lagomarsino, R. Landua, E. Lodi Rizzini, M. Macrì, N. Madsen, D. Mitchard, P. Montagna, A. Rotondi, G. Testera, A. Variola, L. Venrurelli, D.P. van der Werf, and Y. Y. A. Collaboration), *Phys. Rev. Lett.*, **95**, 025002 (2005).
9. H. Higaki, N. Kuroda, K. Yoshiki Franzen, Z. Wang, M. Hori, A. Mohri, K. Komaki, and Y. Yamazaki, *Phys. Rev. E*, **70**, 026501 (2004).
10. G. Andresen, W. Bertsche, P. Bowe, C. Bray, E. Butler, C. Cesar, S. Chapman, M. Charlton, J. Fajans, M. Fujiwara, R. Funakoshi, D. Gill, J. Hangst, W. Hardy, R. Hayano, M. Hayden, R. Hydomako, M. Jenkins, L. Jørgensen, L. Kurchaninov, R. Lambo, N. Madsen, P. Nolan, K. Olchanski, A. Olin, A. Povilus, P. Pusa, F. Robicheaux, E. Sarid, S. S. E. Nasr, D. Silveira, J. Storey, R. Thompson, D. van der Werf, J. Wurtele, and Y. Yamazaki, *Phys. Rev. Lett.*, **100**, 203401 (2008).
11. H. Knudsen, H.-P. Kristiansen, H. Thomsen, U. Uggerhøj, T. Ichioka, S. Møller, C. Hunniford, R. McCullough, M. Charlton, N. Kuroda, Y. Nagata, H.A. Torii, Y. Yamazaki, H. Imao, H. Andersen, and K. Tökesi, *Phys. Rev. Lett.*, **101**, 043291 (2008).

Radial Compression of a Non-neutral Plasma in a Non-uniform Magnetic Field of a Cusp Trap

H. Saitoh[*,†], A. Mohri[*], Y. Enomoto[*,**], Y. Kanai[*] and Y. Yamazaki[*,**]

[*]*Atomic Physics Laboratory, RIKEN, Wako 351-0198, Japan*
[†]*Graduate School of Frontier Sciences, University of Tokyo, Kashiwa 277-8561, Japan*
[**]*Graduate School of Arts and Sciences, University of Tokyo, Komaba 153-8902, Japan*

Abstract. Spectroscopic comparison of antihydrogen and hydrogen atoms is one of the best candidates for the stringent tests of the CPT symmetry, and intensive studies are being carried out by using Antiproton Decelerator at CERN. The ASACUSA collaboration has constructed a superconducting cusp trap for the formation, trapping and extraction of antihydrogen atoms, where a quadrupole magnetic field is generated by a pair of anti-Helmholtz coils with anti-parallel currents. The cusp configuration is considerably advantageous for the extraction of spin-polarized and ground-state antihydrogen beams that are ideal for the spectroscopic measurements of hyperfine structures of the ground state of antihydrogen. For the effective generation of antihydrogen atoms, it is essential to form high density and stable plasmas of antiproton and positrons. In this study, we applied a rotating electric field to an electron plasma in the inhomogeneous cusp magnetic field, and demonstrated the effective radial compression of a non-neutral plasma in a broad frequency range. The compression rate depended on the rotating frequency and had a broad peak extending on both sides of a longitudinal (1,0) mode frequency, which was the only observed characteristic frequency. The similar procedure can in principle be applied to positron and antiproton plasmas, and the results are one of necessary steps toward antihydrogen experiments in the cusp trap.

Keywords: non-neutral plasma, pure electron plasma, cusp magnetic field, antihydrogen synthesis
PACS: 37.10.-x, 52.27.Jt, 36.10.-k, 52.27.Aj

INTRODUCTION

Compression techniques of charged particles have a wide range of scientific applications in the fields of non-neutral plasma, atomic, and particle physics. Recently, experiments on antihydrogen synthesis are being carried out intensively by using the Antiproton Decelerator (AD) [1] at CERN. Spectroscopic comparison of antihydrogen atoms with hydrogen ones provides one of the stringent tests of the CPT symmetry, and so far two groups have succeeded the production of cold antihydrogen atoms [2, 3]. For efficient production of antihydrogen atoms in these experiments, a method to form high density antiproton and positron plasmas is one of the essential issues.

In the antihydrogen experiments carried out at CERN, 5.3 MeV antiprotons from the AD were decelerated, trapped, and mixed with positron plasmas to form antihydrogen atoms via recombination processes in uniform magnetic fields of "nested" Penning traps. In a trap with a uniform magnetic field, however, the generated neutral atoms feel no attracting or repulsive forces, and therefore it is not straightforward to manipulate antihydrogen atoms for further spectroscopic studies. For future experiments on high-precision laser or microwave spectroscopy of antihydrogen aiming for the CPT symmetry test, it is required that the generated antihydrogen atoms in high Rydberg states are trapped for a time long enough for cascade down to their ground state [11, 12, 13]. Antihydrogen atoms with a magnetic moment μ undergo $-\mu\nabla|\mathbf{B}|$ force in an inhomogeneous magnetic field **B**, and thus cold antihydrogen atoms in the low-field seeking states can in principle be confined in minimum-B magnetic field configurations [14, 15].

Several approaches are being made to introduce a magnetic gradient on a trap geometry and to simultaneously confine both neutral hydrogen atoms and non-neutral plasmas. In the standard Ioffe-Pritchard trap [16], the quadrupole field of Ioffe bars breaks the axial symmetry of a magnetic configuration, which may lead to instability or rapid losses of non-neutral plasmas [17]. A new experiment was proposed for antihydrogen spectroscopy by using a trap with octopole magnetic field for reducing the field asymmetry and resultant perturbations, and both confinement and compression of antimatter plasmas were successfully demonstrated without significant diffusive loss [18].

In the ASACUSA collaboration, the use of a cusp magnetic field configuration (an axisymmetric magnetic quadrupole) has been proposed for the scheme to synthesize and trap cold antihydrogen atoms [19, 20, 21, 22, 12]. When antiproton and positron clouds are mixed near the magnetic null point of the cusp trap, cold antihydrogen atoms are synthesized and extracted as ultra-slow spin-polarized beams. In this study, we have carried out experiments on electron plasmas in a spindle cusp region in order to understand the basic properties of non-neutral plasmas in the cusp field. We applied a rotating wall technique to the cusp non-neutral plasma and found a way to make an effective radial compression. Although the application of the RF field of the rotating wall heated up the plasma, the synchrotron radiation in the strong cusp magnetic field quickly cooled the plasma, and as a result a cold and high density plasma was successfully formed [21].

EXPERIMENTAL SETUP

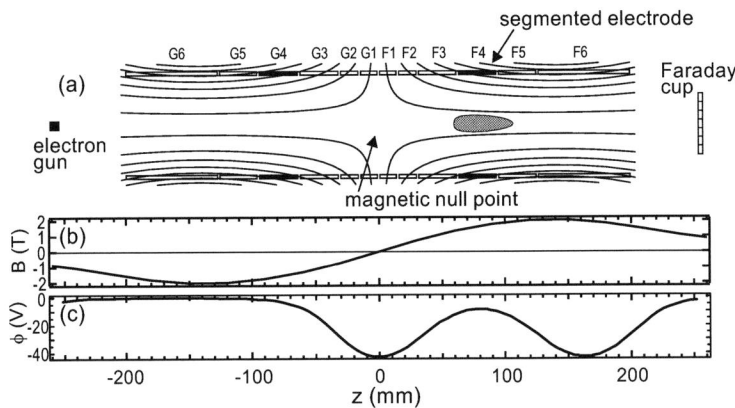

FIGURE 1. (a) Schematic view of the trap electrodes and typical (b) magnetic and (c) electrostatic potential profiles on the device axis $r = 0$.

Schematic view of the trap and field profiles of the superconducting cusp device [21] is shown in Fig. 1. The trap configuration is generated by the combination of a cusp magnetic field, expressed by a vector potential $A_\theta(r,z) = (B_0/L)rz$ in the cylindrical coordinates (r,θ,z), and an electrostatic potential. The "throat" of the field lines is located at $z = \pm 150$ mm, where the maximum field strength on the device axis is 2.0 T at the coil current of 60 A. In the confinement region of the trap, the typical field strength in this experiment was 1.6 T at $z = \pm 80$ mm. The variation of the field strength at $z = 40 - 120$ mm was 55%. The trap consists of twelve cylindrical electrodes of the inner diameter of 80 mm. In the trap region, a harmonic electrostatic potential $\phi_V(r,z) = -\phi_0 \left[(z-L)^2 - r^2\right]/L^2$ was applied. The trap system was evacuated to the base pressure of 2×10^{-7} Pa. As shown in Fig. 1 (c), the potential well for electron trap was located near $z = L = +80$ mm. Here ϕ_0/L^2 was set in the range from 3.6×10^3 to 1.1×10^4 V/m^2 in the experiment. In the most often used case, $\phi_0/L^2 = 9.0 \times 10^3$ V/m^2 and the depth of the potential well was 30 V. The longitudinal bounce motion of an electron in the potential well is approximated as a harmonic oscillation of frequency $f_b = \sqrt{e\phi_0/2\pi^2 m_e L^2} = 9.0$ MHz. In Fig. 1 (a), F4 is a segmented electrode that was used for applying an azimuthally rotating $m = 1$ electric field to the plasma. Sinusoidal RF on the four electrode segments had phase differences of $\pi/2$ in the direction parallel or opposite to the $\mathbf{E} \times \mathbf{B}$ plasma rotation and f_{RW} was < 15 MHz.

Electrons were injected by an electron gun as a train of pulsed beams. In the injection phase, the potential wall on the electron gun side was lowered synchronously to the arrival of each beam pulse and thereby electrons were trapped inside the potential well. Electrons were dumped on a Faraday cup located at the downstream side of the trap (see Fig. 1 (a)). The Faraday cup is coaxially segmented into eight with the radial pitch of 4 mm and the total diameter is 60 mm, which allows to monitor the line-integrated density n_l (integration of density along magnetic field lines) of the electron plasma. We also measured the parallel electron temperature by shallowing the potential wall and analyzing the energy profiles of escaping electrons [23].

EXPERIMENTAL RESULTS AND DISCUSSION

Excitation of characteristic modes

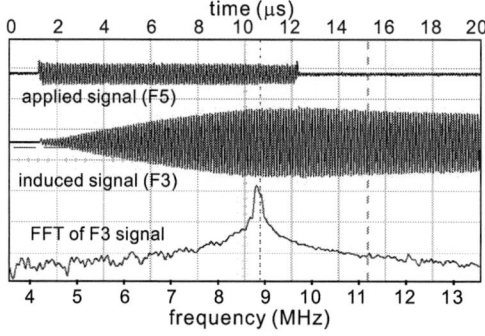

FIGURE 2. Response of plasma on external fluctuations. Applied RF frequency was 8.9 MHz, in resonance with the characteristic plasma frequency f_z.

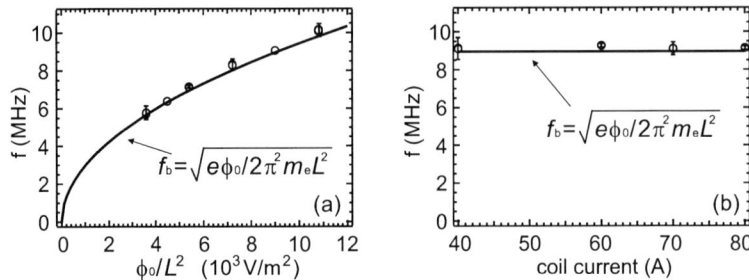

FIGURE 3. Observed resonant frequencies f_z as functions of (a) gradient of the external electrostatic potential well, represented by ϕ_0/L^2, and (b) cusp magnetic field strength.

Figure 2 shows the excitation of plasma oscillation modes. An external RF was applied in the longitudinal direction of the device by using F5. When the applied RF resonated with a mode of frequency f_z, the amplitude of the excited plasma oscillation grew almost linearly, as shown in Fig. 2. This mode was also excited in an off-resonant condition if the RF input was large enough. By sweeping the frequency of sinusoidal wave on F5 from 10 kHz to 30 MHz, we found only one resonant mode corresponding to the center of mass oscillation of an electron plasma: $f_z = f_b$. Figure 3 shows the frequencies of the observed oscillation mode f_z as functions of ϕ_0 and the coil current being proportional to the magnetic field strength B. The mode frequency had a dependence of $f_z \propto E^{1/2} B^0$, where E is the external electric field strength. The solid curve f_b in Fig. 3 well reproduces the observed mode frequencies. Although the electron temperature varied during the confinement cycle, as discussed later, the observed frequency stayed constant [24].

Radial compression of cusp plasmas by using a rotating wall

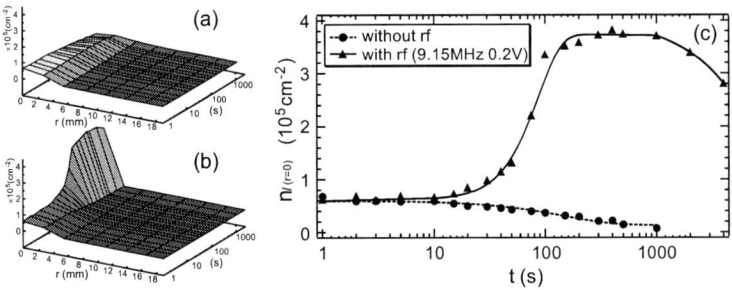

FIGURE 4. Temporal evolution of the line-integrated density profiles of a plasma (a) with and (b) without the rotating wall, and (c) n_l: line-integrated density at $r = 0$. The solid and dotted lines are to guide the eyes only.

Temporal evolutions of n_l profiles were measured by the Faraday cup as displayed in Fig. 4. When the rotating wall was not applied (Fig. 4 (a), circles in (c)), the plasma radially expanded, and after $t \sim 400$ s, the total electron number decreased with a $1/e$

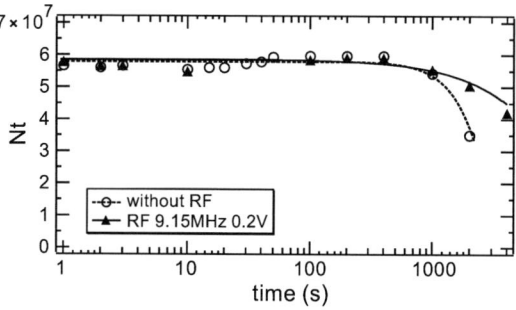

FIGURE 5. Temporal evolutions of total trapped charge with and without application of rotating RF.

time constant of 3000 s as shown in the circles in Fig. 5. The line-integrated density at $r = 0$ was $n_{l(r=0)} = 6.7 \times 10^4$ cm^{-2} at $t = 1$ s, and decreased to $n_{l(r=0)} = 7.2 \times 10^3$ cm^{-2} at $t = 1000$ s with a time constant of $\tau_E = 150$ s.

In contrast, by applying the rotating wall, effective compression was observed as shown in Fig. 4 (b). The rotating wall was applied in the direction of the $\mathbf{E} \times \mathbf{B}$ rotation of the plasma at the frequency of $f_{RW} = 9.15$ MHz, which was 0.25 MHz higher than f_z. The amplitude of the applied RF was $V_{RW} = 0.2$ V. The solid triangles with the fitting curve in Fig. 4 (c) show that n_l increased and it saturated at $t \sim 400$ s, where $n_{l(r=0)} = 3.8 \times 10^5$ cm^{-2}. After the application of the rotating wall for 400 s, 74 % of the initial electrons were radially compressed into the central region ($r < 1.4$ mm). The total number and $n_{l(r=0)}$ were both unchanged for 1000 s after reaching the peaky density profile (Fig. 5).

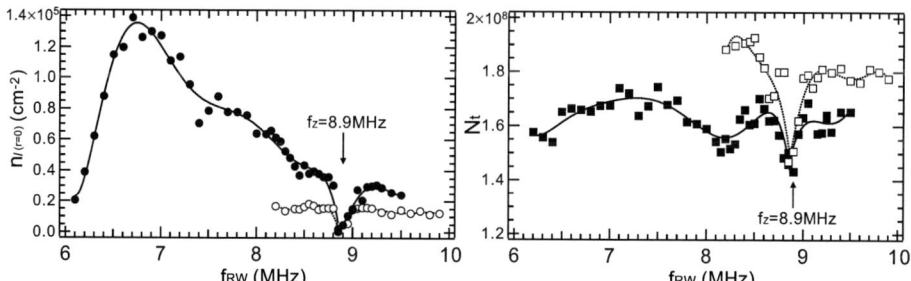

FIGURE 6. Total electron number N_t and line-integrated density on the device axis $n_{l(r=0)}$ at $t = 20$ s as a function of applied rotating wall frequency. Directions of the applied rotating wall were equal to (solid circles and solid squares) and opposite to (open circles and open squares) the $\mathbf{E} \times \mathbf{B}$ plasma rotation.

In Fig. 6, the central line-integrated density $n_{l(r=0)}$ and the total electron number N_t were measured after the application of the rotating wall for 20 s. When the rotating wall was not applied, $n_{l(r=0)} = 2.1 \times 10^4$ cm^{-2}. When the rotating wall was applied in the $\mathbf{E} \times \mathbf{B}$ rotation direction, radial compression was observed in a broad f_{RW} range from approximately 6 MHz to 9.5 MHz except for frequencies from 8.85 MHz to 9.0 MHz. When f_{RW} was close to the observed longitudinal resonance mode frequency of $f_z = 8.9$ MHz, both the numbers of electrons in the central region and the total electrons decreased quickly regardless of the rotation direction.

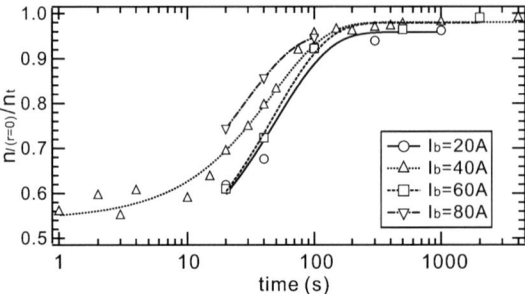

FIGURE 7. Central plasma density normalized by the average density for four different coil currents. Frequency of the applied rotating wall RF was 9.15 MHz

Figure 7 shows the temporal evolutions of the central plasma density (normalized by the average density) for four different coil currents. The rotating wall RF with an amplitude of 0.6 V was applied for 20 s. The compression effects were similarly observed in the all applied coil current values, and there was no strong dependence on B. The typical time constant of the compression was of the order of ~ 100 s.

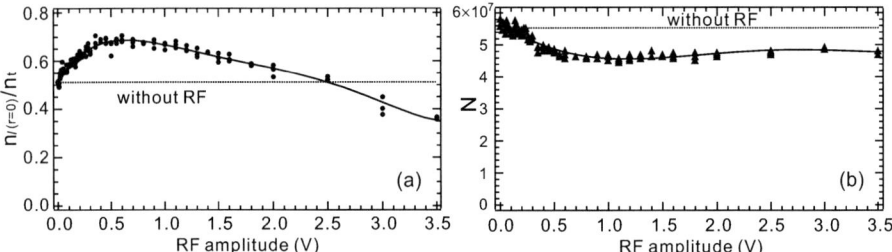

FIGURE 8. (a) Compression effect and (b) total electron number for different rotating RF amplitude. Rotating wall RF of 9.15 MHz was applied for 20 s.

Figure 8 shows the central plasma density for different RF amplitude V_{RF} of the rotating wall. When the RF was not applied, $n_{l(r=0)}/n_t = 0.51$ after 20 s of confinement. The compression efficiency had a broad peak around $V_{RF} = 0.6$ V, and as V_{RF} was further increased, effective compression was not observed above $V_{RF} = 2.5$ V. As described later, application of the rotating wall heats the plasma, but cyclotron radiation cooling prevents the increase of the temperature.

Electron temperature

As shown in Fig. 9, temporal evolutions of the electron temperature parallel to **B**, $T_{e\|}$, was measured by using the energy distribution of electrons leaked to the Faraday cup. In the initial phase of confinement at $t = 0.2$ s, $T_{e\|} = 5.7$ eV. When the rotating wall field was not applied, it cooled quickly and reached 0.4 eV in $t = 10$ s and after then stayed almost constant. Such a steady state is also another side-evidence on the existence of a rigid-rotor equilibrium of a non-neutral plasma in the spindle cusp configuration. The observed $1/e$ cooling time of $T_{e\parallel}$ was 2.5 s, which is comparable to the typical cooling time caused by the synchrotron radiation in a magnetic field given by the Larmor formula for the typical field strength $B = 1.6$ T: $\tau_D \sim 3\pi\varepsilon_0 m_e c^3/e^2 \omega_{ec}^2 \sim 1.0$ s. Although the variation of field strength was small, quicker cooling was observed for stronger B as shown in Fig. 9 (b). The parallel electron temperature did not decrease lower than 0.4 eV, which could be due to the effects of the warm bore and external noise. Further study will be carried out in the near future using a cryogenic cusp trap.

The closed triangles in Fig. 9 (a) show the temperature evolution when the rotating wall was turned on, which revealed that it reached an equilibrium of $T_{e\parallel} = 1.7$ eV, four times higher than that without the rotating wall. When the RF was turned off, $T_{e\parallel}$ dropped to the same temperature 0.4 eV as in the case of no RF application, and this cooling rate was comparable to τ_D as shown in Fig. 9 (b). The time scale of the radial expansion of the compressed plasma $\tau_E \sim 150$ s is much longer than τ_D, and the rotating wall can significantly compress the plasma practically keeping the plasma temperature unchanged.

In future experiments, a positron plasma in one side of a spindle cusp will be transported to the opposite side in order to induce recombinations with antiprotons to form antihydrogen atoms. In this process, particles located near the center axis of the device are primarily transferred through the magnetic null point, and the plasma may take unstable hollow structures. The rotating wall can be utilized for the stable transfer of positrons by the radial compression that compensate the decrease of particles on the device axis.

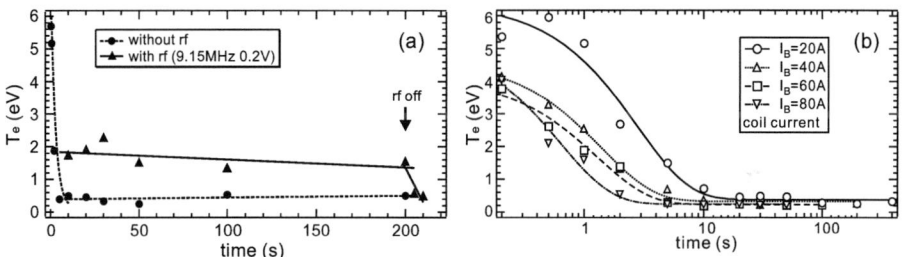

FIGURE 9. Temporal evolutions of $T_{e\parallel}$ (a) when the rotating wall was applied from $t = 0$ to 200 s (triangles) and not applied (circles) and (b) in the variation of field strength without the application of rotating wall.

CONCLUSION

To summarize, we have demonstrated that a single-component plasma in the inhomogeneous cusp magnetic field can be effectively compressed by using the rotating wall technique as in the case of a plasma in a uniform magnetic field. The compression was observed with a broad frequency range of a rotating wall in the $\mathbf{E} \times \mathbf{B}$ velocity direction except for frequencies close to the longitudinal resonance frequency of the plasma, which is the only observed mode of a spindle cusp electron plasma. The compression and cooling of an electron plasma were simultaneously realized. Although the present experiment was carried out using an electron plasma, the same procedure can in principle be applied to a positron plasma, which is particularly important for the formation and extraction of antihydrogen atoms.

ACKNOWLEDGMENTS

This work was funded by Special Research Projects for Basic Science of RIKEN. The work of H. S. was supported in part by the Special Postdoctoral Researchers Program of RIKEN.

REFERENCES

1. S. Maury, Hyperfine Interact., **109**, 43 (1997).
2. M. Amoretti *et al.*, Nature **419**, 456 (2002).
3. G. Gabrielse *et al.*, Phys. Rev. Lett. **89**, 213401 (2002).
4. X.-P. Huang, F. Anderegg, E. M. Hollmann, C. F. Driscoll, and T. M. O'Neil, Phys. Rev. Lett. **78**, 875 (1997).
5. X.-P. Huang, J. J. Bollinger, T. B. Mitchell, and W. M. Itano, Phys. Rev. Lett. **80**, 73 (1998).
6. F. Anderegg, E. M. Hollmann, C. F. Driscoll, Phys. Rev. Lett. **81**, 4875 (1998).
7. E. M. Hollmann, F. Anderegg, C. F. Driscoll, Phys. Plasmas **7**, 2776 (2000).
8. Y. Soga, Y. Kiwamoto, and N. Hashizume, Phys. Plasmas **13**, 052105 (2006).
9. J. R. Danielson and C. M. Surko, Phys. Rev. Lett. **94**, 035001 (2005); J. R. Danielson and C. M. Surko, Phys. Plasmas **13** 055706 (2006).
10. R. G. Greaves and C. M. Surko, Phys. Rev. Lett. **85** 1883 (2000); R. G. Greaves and C. M. Surko, Phys. Plasmas **8**, 1879 (2001).
11. Y. Yamazaki, Physica Scripta **T110**, 286 (2004).
12. T. Pohl, H. R. Sadeghpour, Y. Nagata, and Y. Yamazaki, Phys. Rev. Lett. **97**, 213001 (2006).
13. W. Bertsche *et al.*, Nucl. Instrum. Methods Phys. Res., Sect. A **521**, 746 (2006).
14. T. M. Squires, P. Yesley, and G. Gabrielse, Phys. Rev. Lett. **86**, 5266 (2001).
15. D. H. Dubin, Phys. Plasmas **8** 4331 (2001).
16. D. E. Pritchard, Phys. Rev. Lett. **51**, 1336 (1983).
17. E. P. Gilson and J. Fajans, Phys. Rev. Lett. **90**, 015001 (2003).
18. G. Andresen *et al.*, Phys. Rev. Lett. **98**, 023402 (2007).
19. A. Mohri *et al.*, Jpn. J. Appl. Phys. **37**, L1553 (1998).
20. A. Mohri and Y. Yamazaki, EuroPhys. Lett. **63**, 207 (2003).
21. A. Mohri *et al.*, in *Physics with Ultra Slow Antiproton Beams* (AIP Conf. Procs. **793**), 147 (2005); A. Mohri *et al.*, in Procs. EPS Conf. Plasma Phys P-4.067 (2006); H. Saitoh *et al.*, Phys. Rev. A **77**, 051403 (2008).
22. N. Kuroda *et al.*, Phys. Rev. Lett. **94**, 023401 (2005); N. Kuroda *et al.*, Phys. Rev. Lett. **100**, 203402 (2008).
23. B. R. Beck, J. Fajans, and J. H. Malmberg, Phys. Rev. Lett. **68**, 317 (1992).
24. H. Higaki and A. Mohri, Jpn. J. Appl. Phys. **36**, 5300 (1997).

Tailored Particle Beams From Single-Component Plasmas

T. R. Weber, J. R. Danielson and C. M. Surko

Department of Physics, University of California, San Diego, La Jolla CA 92093-0319

Abstract. Recently, we developed a non-destructive technique to create narrow beams of electrons (or positrons) of adjustable width and brightness from single-component plasmas confined in a Penning-Malmberg trap [Weber et al., Phys. Plasmas **13**, 123502 (2008)]. Here, we review highlights of that work and discuss the distributions in energy of the extracted beams. A simple model for beam extraction predicts Gaussian beam profiles, with transverse spatial widths dependent on the number of particles in the beam. A Maxwellian energy distribution is predicted for small beams. The predictions of the theory are confirmed using electron plasmas. Extraction of over 50% of a trapped plasma into a train of nearly identical beams is demonstrated. Finally, the possibility of creating high quality, electrostatic beams by extraction from the confining magnetic field is discussed.

Keywords: bright beams, narrow beams, positrons, antiprotons, antimatter, nonneutral plasmas.
PACS: 52.27.Jt, 52.35.-g, 34.80.Uv, 41.75.Fr, 41.85.Ar.

INTRODUCTION

Charged particle beams are used for a wide range of applications in science and technology. In the case of electrons, beams generated by a simple heated cathode or field emission are adequate for most applications. However, when the particles are more difficult to obtain, as is the case with positrons and antiprotons, for example, more refined techniques are required. In the case of antimatter, it has proven convenient to use trap-based beams, where the particles are first accumulated efficiently and cooled in an electromagnetic trap, then a beam or pulse of particles is extracted [1-4]. In this paper, we describe the use of a Penning-Malmberg trap to create high-quality, trap based beams, having in mind that this is directly relevant to the creation of state-of-the-art positron beams. These beams are extracted from the center of the single-component plasma by carefully lowering an end-gate confining potential. Much of this paper is an overview of recent work that is described in more detail elsewhere [5, 6].

Many applications require positron or antiproton beams with small transverse spatial extent and small energy spreads [4, 7-9]. We describe here techniques to create such beams with radii, ρ_b, as small as 10 μm and energy spreads of the order of the temperature of the parent plasma (i.e., which is cooled by cyclotron radiation in the confining, 5 T magnetic field). For rare particles, such as positrons, it is advantageous to use them efficiently. As discussed below, that can be done with near 100% efficiency using the techniques described here.

Penning-Malmberg Trap for Beam Extraction

The plasma particles are accumulated and stored in a Penning-Malmberg trap, such as that shown schematically in Fig. 1 (a) [6]. It consists of a set of cylindrical electrodes, of radius $R_W = 1.2$ cm, in a uniform magnetic field of strength $B = 5$ T. The particles are confined radially by the magnetic field and axially by voltages, V_C applied to electrodes at each end. The resulting plasma is in thermal equilibrium at temperature T. The plasma is a uniform-density rigid rotor, rotating at an $E \times B$ frequency, $f_E = cn_0e/B$, proportional to the equilibrium density, n_0. The plasma parameters are z-independent, thus making r and θ the coordinates of interest.

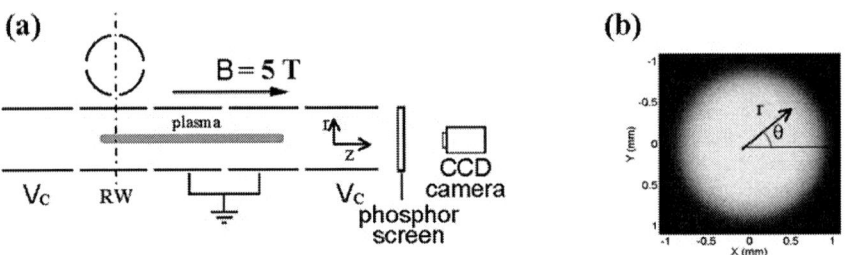

FIGURE 1. (a) Schematic diagram of the experimental arrangement; and (b) a CCD images of the areal plasma density, $\sigma_z(r, \theta)$, for an equilibrium, "flat-top" plasma.

The principal diagnostic used here is imaging the two-dimensional (i.e., areal) plasma density using a phosphor screen located outside the trap and a CCD camera. By quickly setting $V_C = 0$ at one end of the plasma, the plasma particles stream out of the trap along the magnetic field. They are accelerated to energies of 5 keV, and then collide with the phosphor screen. The resulting fluorescent light is imaged with a CCD camera to obtain the z-integrated, areal plasma density profile, $\sigma_z(r,\theta)$. A typical CCD image is shown in Fig. 1 (b). The plasma density, n, is then $n(r, \theta) = \sigma_z /L_P$, where L_P is the plasma length.

Another important technique for the work described here is use of rotating electric fields to compress the plasma radially [i.e., the so-called "rotating wall" (RW) technique] [10]. The segmented electrode for this is shown in Fig. 1 (a). Phased, sinusoidal voltages are applied to each of the four segments. This azimuthally asymmetric potential rotates around the z-axis, hence the name "rotating wall". For sufficiently large sine wave voltages, the plasma spins up (or down) to the applied RW frequency, f_{RW} [10]. Because f_E is proportional to the equilibrium density, n_0, the plasma density can thus be set, *in situ*, in a non-destructive way, by simply tuning f_{RW}.

BEAM EXTRACTION

To extract a beam from a trapped plasma, the confining potential, V_C, at one end of the plasma is lowered to a value V_E for about 15 μsec. This time is chosen to be sufficiently long so that particles with sufficient energy have enough time to escape, but short enough so that instabilities and radial transport are negligible. This process is illustrated schematically in Fig. 2 (a). Because the plasma potential is highest at the (radial) center of the plasma, the beam is composed of particles from this region. This is highly beneficial, as it creates beams with spatial extent much smaller than that of the parent plasma. Fig. 2 (b) shows CCD images of a plasma before and after a beam is extracted. Notice the small hole at the center, illustrating the location of the particles that exited the trap.

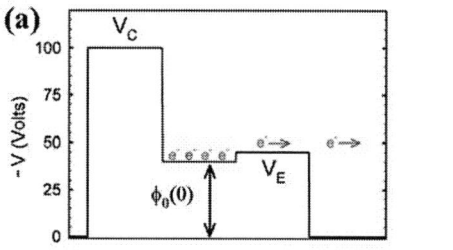 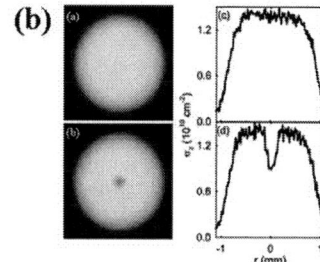

FIGURE 2. (a) Schematic diagram of the beam extraction process. (b) CCD camera images of the areal plasma density, profile, $\sigma_z(r, \theta)$ for a "flat top" plasma before, and 10 μs after, beam extraction; also shown are the corresponding (slice) distributions, $\sigma_z(r)$.

THEORETICAL CONSIDERATIONS

When the end-gate potential is lowered to V_C, particles will begin to escape. As they do, the space-charge potential near the center of the plasma will change in such a way as to inhibit further particles from leaving. The condition that particles escape will be

$$E \geq E_{MIN} = -e(V_E - \phi_0(r) + \Delta\phi(r)), \quad (1)$$

where E_{MIN} is the energy required for a particle to escape at the end of the beam extraction, $\phi_0(r)$ is the equilibrium plasma potential and hence is quadratic in r, and $\Delta\varphi(r)$ is the change in the plasma potential due to the extracted beam particles. After extraction, the new density and space-charge potential profiles are $n(r) = n_0(r) - \Delta n(r)$, and $\varphi(r) = \varphi_0(r) - \Delta\varphi(r)$.

The areal density profile, $\sigma_b(r)$, as measured using the phosphor screen, will be $\sigma_b(r) = \Delta n(r) \times L_P$. Assuming an initially uniform density plasma in thermal equilibrium, $\sigma_b(r)$ can be written

$$\sigma_b(r) = 2n_0 L_P \int_{E_{MIN}}^{\infty} f(E_{\parallel})dE_{\parallel} = 2L_P n_0 erfc\left(\sqrt{-\frac{e}{T}(V_E - \phi_0(r) + \Delta\phi_0(r))}\right). \quad (2)$$

Assuming E_{MIN} is on the tail of the Maxwellian, $erfc(x) \sim exp(-x^2)/x$. Then approximating $\Delta\varphi(r)$ as the change in potential due to a "flat top" beam of width, ρ_b, and using the expression for $\varphi_0(r)$ [6],

$$\sigma_b(r) = \sigma_{b0} \exp\left[-\left(\frac{r}{2\lambda_D}\right)^2 + \xi\left(\frac{r}{\rho_b}\right)^2\right], \quad \text{where} \quad \xi = \frac{e^2 N_b}{L_p T}. \quad (3)$$

Here, σ_{b0} is a function of V_E and approximately constant for all r. Defining ρ_b as the half width to 1/e, this width becomes,

$$\rho_b = 2\lambda_D (1+\xi)^{1/2} \quad (4)$$

The quantity ξ is the key parameter that determines the size of extracted beams. For small values of ξ, Eq. (3) predicts a Gaussian beam with a full width to 1/e of $4\lambda_D$.

A key quantity of interest is the distribution of beam particles in parallel energy space, $f(E_\parallel)$, which is defined as dN_b/dV_E, evaluated at $V_E = E_\parallel$. From Eq. 3, it can be seen that $\sigma_{b0} \sim N_b$. In the limit of small beams (i.e., $\xi \ll 1$), $e\Delta\varphi/T$ can be neglected in Eq. (2), so that

$$\left.\frac{d\sigma_{bo}}{dV_E}\right|_{V_E = E_\parallel} \propto f(E_\parallel) \propto \frac{\exp[-(E_\parallel - e\phi_0(0))/T]}{\sqrt{(E_\parallel - e\phi_0(0))/T}}. \quad (5)$$

Equation (6) is just the tail of the energy distribution of a Maxwellian sitting at an electrical potential of $\varphi_0(0)$. Thus, for small beams, the parallel energy spread is $\sim T$.

SINGLE BEAMS

Shown in Fig. 3 (a) are areal density profiles for beams of different amplitudes. The corresponding beam widths as a function of ξ are shown in Fig. 3 (b). The results of the previous section predict Gaussian beam profiles for small beams with a full width

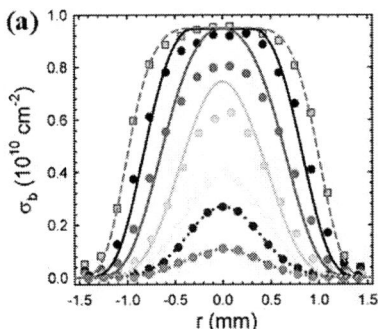
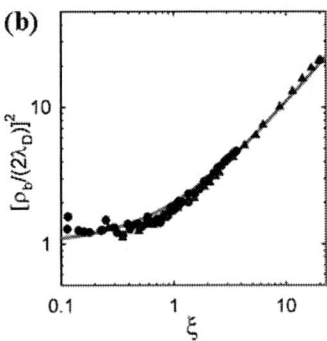

FIGURE 3. (a): Profiles, $\sigma_b(r)$, of extracted beams from a $T = 1.0$ eV plasma with density, $n_0 \approx 1 \times 10^9$ cm^{-3}. From smallest to largest, $\xi \approx 0.1, 0.3, 0.5, 1.0, 1.9, 2.8$ for each beam. The three smallest beams are fit ($\cdots$) to Eq. 3. The initial plasma profile, $\sigma_z(r)$, (■) is also shown. (b): Beam width parameter, ρ_b, plotted vs. ξ for T = 1.0 eV (●), and 0.2 eV (▲). The prediction (—) from Eq. 4, with no fitted parameters, is also shown.

to 1/e of $4\lambda_D$, and the data in Fig 3 are in agreement with this prediction. The predictions of Eq. 4 are valid for all values of ξ tested, while the Gaussian profile predicted in Eq. 3 is valid for $\xi < 1$ (c.f., Fig. 3). The solid lines in Fig. 3 (a) are the results of numerical calculations for larger beams; see Ref. [6] for details.

Shown in Fig. 4 are the radial profiles of beams extracted before and after RW compression of the parent plasma. These data illustrate the ability to create, nondestructively, beams with narrow transverse spatial extent and thus enhanced brightness. For the beam shown in Fig 4 (b), the beam brightness was increased by a factor of 40, while ξ remains constant.

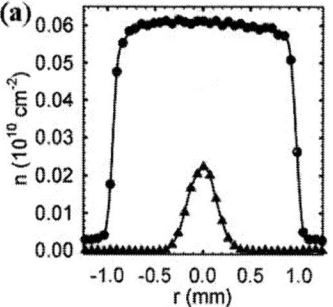 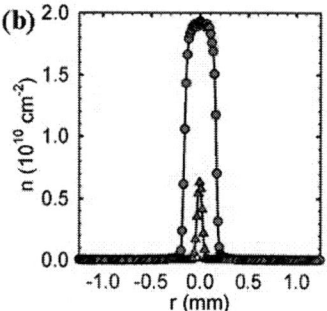

FIGURE 4. Beams with $\xi = 0.5$, extracted from a plasma (a) before RW compression, with $n_0 \approx 0.6 \times 10^9$ cm^{-3}; and (b) after RW compression, with $n_0 \approx 20 \times 10^9$ cm^{-3}. In both cases, $T = 0.05$ eV.

The parallel energy distributions of two extracted beams are shown in Fig. 5. As shown in Fig. 5 (a), for the smallest beam, with $\xi = 0.02$, the predictions of Eq. (5) are in excellent agreement with the data. However, as shown in Fig. 5 (b), when ξ is increased to the seemingly still small value of 0.1, the predictions of Eq. (5) deviate significantly from the data. The resulting beam energy distributions for larger beams, the shifts in mean beam energy and energy spread are due to the non-negligible effects of $\Delta\varphi(r)$ and will be discussed elsewhere.

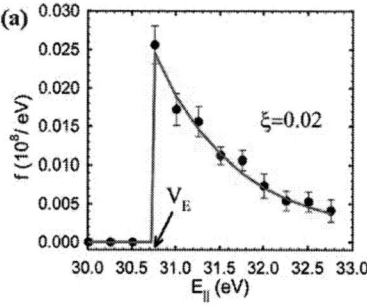 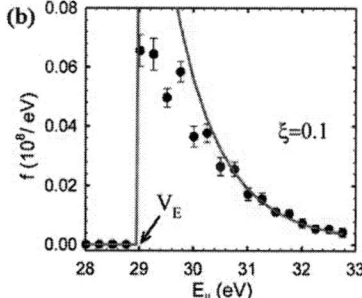

FIGURE 5. Energy distributions of beams extracted from a plasma with $T = 1.0$ eV, $n_0 \approx 1 \times 10^9$ cm^{-3} for (a) $\xi = 0.02$, and (b) $\xi = 0.1$. Also shown are the predictions of Eq. 6 (—), for $\varphi_0(0) = 28$ V.

EXTRACTION OF MULTIPLE BEAMS

Due to the relative scarcity of positrons, it is desirable to use them as efficiently as possible. Data for such an efficient extraction from an electron plasma are shown in Fig. 6 where 20 nearly identical beams were extracted utilizing over 50 % of the parent plasma with essentially no loss of particles. The beams were extracted serially, waiting for the plasma to recover to its equilibrium state [i.e., from a state such as that shown in Fig. 2 (b)], then pulsing V_C to another value of V_E which was carefully chosen to maintain constant N_b. The plasma recovers in ~ 0.5 ms via an unstable diocotron mode [11]. Care was taken to ensure that the plasma relaxes to thermal equilibrium, which is expected to occur on a shorter time scale (i.e., τ_{ee} < 0.5 msec) [6]. During extractions, the RW is kept on to maintain constant n_0, thus fixing λ_D and ρ_b. The nearly identical nature of the radial beam profiles is illustrated in Fig. 6 (b).

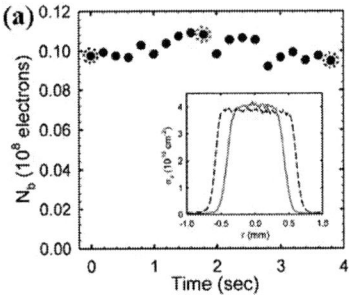 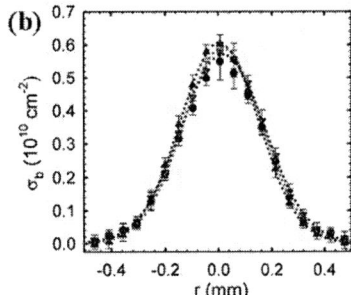

FIGURE 6. (a): N_b for twenty beams extracted consecutively from the same plasma with $T \approx 0.3$ eV and $n_0 \approx 2 \times 10^9$ cm^{-3}. Here, $<N_b>$ = 0.1 ± 0.005 x 10^8 and $\xi \approx 0.2$. The inset shows the radial profiles, $\sigma_z(r)$, of the plasma before (---) and after (–) extraction of the 20 beams. (b) radial beam profiles, $\sigma_b(r)$, for the 1st (●), 10th (▲), and 20th (▼) beam pulse.

ELECTROSTATIC BEAMS

For some applications, pulses of positrons are desired in a magnetic-field-free region (e.g., to accommodate electrostatic focusing techniques) [12]. This can be accomplished by pulling the particles out of the field in a non-adiabatic manner. However, for the 5 T field used here, this would require the magnetic field to drop to zero in distances < 100 μm, which would be difficult to achieve experimentally. Thus we propose extracting the particles from the field in two stages: Stage I, a slow adiabatic reduction of B along the beam path from 5 T to 0.5 mT; then Stage II, a rapid non-adiabatic extraction from B = 0.5 mT to zero. In Stage I, the field will be decreased over large distances so that the particles remain on field lines while the adiabatic invariant, E⊥/|B| is conserved. This has the effect of cooling the perpendicular temperature, $T⊥$, while the beam expands spatially due to the diverging field lines. To conserve energy, $T_∥$ increases accordingly.

In the non-adiabatic stage, the particles are removed from the field lines, receiving an impulse in the azimuthal direction to conserve canonical angular momentum. This results in an increase in $E_\perp$ with a corresponding decrease in $E_\parallel$. This kick is dependent upon r, so the finite spatial extent of the beam results in an increased spread in $T_\perp$ and $T_\parallel$. Table I summarizes the predicted beam parameters for a parent plasma with $n_0 \approx 1 \times 10^{10}$ cm^{-3} and $T = 30$ K (i.e., ≈ 3 meV) that could be created with cryogenically cooled electrodes. While the beam width has increased by a factor of 100, and $T_\perp$ by a factor of 3, such a positron beam would be of excellent quality for a variety of applications, including positron-atomic physics and the characterization of materials.

Table I. Parameters for the creation of positron pulses in a magnetic, field-free region, starting with (I) an initial pulse in a 5 T field; (II) adiabatic transfer to a field of 0.5 mT; and then (III) non-adiabatic extraction from the field.

PARAMETER	STAGE I	STAGE II	STAGE III
B (Tesla)	5	5×10^{-4}	0
$T_\perp$ (meV)	3	6	9
$T_\parallel$ (meV)	3	3×10^{-4}	3
ρ_b (cm)	5×10^{-4}	5×10^{-2}	5×10^{-2}

CONCLUSION

Described here is a method to extract beams of tailored width and brightness in a non-destructive, reproducible manner from plasmas in a Penning-Malmberg trap. A formalism is described that predicts the beam width and energy spread, namely the key parameters of interest for a range of applications. The ability to extract multiple, nearly identical beams is demonstrated, utilizing over 50 % of a single trapped plasma with no loss of particles. Finally, a scenario is discussed in which the techniques described here can be used to produce high-quality electrostatic beams that are expected to be useful for a variety of positron applications.

ACKNOWLEDGMENTS

We wish to acknowledge helpful conversations with R. G. Greaves and the expert technical assistance of E. A. Jerzewski. This work was supported by the NSF, Grant Nos. PHY-03-54653 and PHY 07-13958.

REFERENCES

1. S. J. Gilbert, C. Kurz, R. G. Greaves, and C. M. Surko, *Appl. Phys. Lett.* **70**, 1944 (1997).
2. C. Kurz, S. J. Gilbert, R. G. Greaves, and C. M. Surko, *Nucl. Instrum. Methods in Phys. Res.* **B143**, 188 (1998).
3. D. B. Cassidy, S. H. M. Deng, R. G. Greaves, and A. P. Mills, Jr., *Rev. Sci. Instrum.* **77**, 073106 (2006).
4. C. M. Surko and R. G. Greaves, *Phys. Plasmas* **11**, 2333 (2004).
5. J. R. Danielson, T. R. Weber, and C. M. Surko, *Appl. Phys. Lett.* **90**, 081503 (2007).
6. T. R. Weber, J. R. Danielson, and C. M. Surko, *Phys. Plasmas* **13**, 123502 (2008).
7. D. B. Cassidy and A. P. Mills, Jr., *Nature* **449**, 195 (2007).
8. M. Amoretti, C. Amsler, G. Bonomi, A. Bouchta, P. Bowe, C. Carraro, C. L. Cesar, M. Charlton, M. Collier, M. Doser, V. Filippini, K. Fine, A. Fontana, M. Fujiwara, R. Funakoshi, P. Genova, J. Hangst, R. Hayano, M. Holzscheiter, L. Jorgensen, V. Lagomarsino, R. Landua, D. Lindelof, E. L. Rizzini, M. Macri, N. Madsen, G. Munuzio, M. Marchesotti, P. Montagna, H. Pruys, C. Regenfus, P. Riedler, J.

8. (cont.) Rochet, A. Rotondi, G. Rouleau, G. Testera, A. Variola, T. Watson, and D. VanderWerf, *Nature* **419**, 456 (2002).
9. G. Gabrielse, N. Bowden, P. Oxley, A. Speck, C. Storry, J. Tan, M. Wessels, D. Grzonka, W. Oelert, G. Schepers, T. Sefzick, J. Walz, H. Pittner, T. Hansch, and E. Hessels, *Phys. Rev. Lett.* **89**, 213401 (2002).
10. J. R. Danielson and C. M. Surko, in *Non-Neutral Plasma Physics Vi*, edited by M. Drewsen, U. Uggerhoj and H. Knudsen (American Institute of Physics Press, 2006), p. 19
11. C. F. Driscoll, *Phys. Rev. Lett.* **64**, 645 (1990).
12. C. M. Surko, G. F. Gribakin, and S. J. Buckman, *J. Phys. B: At. Mol. Opt. Phys.* **38**, R57 (2005).

Investigations on Cooling Mechanisms of Highly Charged Ions at HITRAP

Giancarlo Maero*, Frank Herfurth*, Oliver Kester*, H.-Jürgen Kluge*, Stephen Koszudowski*, Wolfgang Quint* and Stefan Schwarz[†]

*Gesellschaft für Schwerionenforschung, D-64291 Darmstadt, Germany
[†]National Superconducting Cyclotron Laboratory, Michigan State University, 48824-1321 East Lansing, USA

Abstract. The upcoming facility HITRAP (Highly Charged Ion TRAP) at GSI will enable high-precision atomic-physics investigations on heavy, highly charged ions at extremely low energies. Species up to U^{92+} will be produced at the GSI accelerator complex by stripping of relativistic ions and injected into the Experimental Storage Ring (ESR) where they are electron-cooled and decelerated to 4 MeV/u. After ejection out of the ESR and further deceleration in a linear decelerator bunches of 10^5 ions will be injected into a Penning trap and cooled to 4 K via electron and resistive cooling. Simulations with a Particle-In-Cell (PIC) code have been carried out to study the dynamics of the ion cloud in the Cooler Trap with focus on resistive cooling in presence of space charge.

Keywords: non-neutral plasmas, highly charged ions, Penning traps, resistive cooling, PIC method.
PACS: 52.27.Jt; 37.10.Ty; 37.10.Rs; 52.65.Rr.

INTRODUCTION

The theory of Quantum Electrodynamics (QED) has shown so far an impressively precise predictive power. Perturbative QED has allowed accurate predictions on observables like the *g-factor* of the free electron or the atomic levels of simple systems, e.g. hydrogen or few-electron atoms. Perturbative QED breaks down for heavy and highly charged ions, where the perturbation term $Z\alpha$ (with Z nuclear charge of the ion and α fine structure constant) becomes ≈ 1 and we enter the *strong field* regime. Non-perturbative theoretical techniques have been developed [1], while experiments with heavy and highly charged ions represent the necessary and perfect workbench for their validation.

The GSI accelerator complex represents a world-unique facility for the production of relativistic beams of heavy and highly charged ions. In order to reach ultimate accuracy the highly charged ions (HCI) must be decelerated. The HITRAP (Highly charged Ion TRAP) facility will serve this purpose [2]. The desired species are created at low charge states, accelerated and partially stripped in the linear accelerator UNILAC. They are further accelerated in the heavy ion synchrotron SIS-18 and then sent through a target foil at energies up to 1 GeV/u, yielding HCI up to U^{92+}. Radioactive species can also be produced by fragmentation or Coulomb dissociation. The beam is successively injected into the Experimental Storage Ring ESR where deceleration and cooling make available 1-μs bunches of some 10^5 HCI at 4 MeV/u every 10 s for injection into the HITRAP facility. Here the beam, rebunched by a Double-Drift Buncher, is further decelerated to 6 keV/u in a decelerator comprising an Interdigital H-type Linac and a Radio Frequency

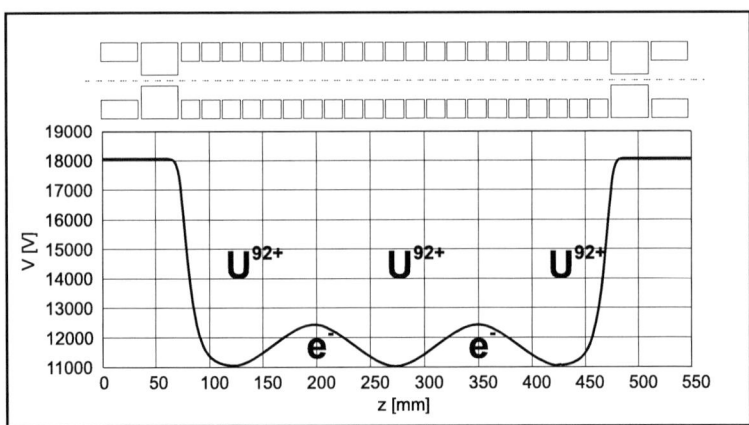

FIGURE 1. Nested trap configuration of the Cooler Trap. In the upper part, a scheme of the electrode stack. The diagram below shows the electric potential (indicative) on the axis at the injection. Notice the inner hills and wells where negative and positive species can be accomodated.

Quadrupole so that it can be captured in-flight by the so-called Cooler Trap. In this Penning trap electron- and resistive cooling of the ion bunch takes place. The cold ion cloud will then be extracted at an energy as low as 4 K and sent to the experimental set-ups.

THE COOLER TRAP: DESIGN AND SETUP

The Cooler Trap is a Penning trap to catch and store 10^5 HCI every 10 s. The ion cloud will be cooled via *electron cooling*, i.e. Coulomb collisions with $10^9 \div 10^{10}$ electrons generated by a photoionization source and confined at the same time in the trap, where they are kept cold via synchrotron radiation in the magnetic field. The thorough study of electron cooling of magnetized plasmas, with particular reference to HITRAP, has been treated by G. Zwicknagel [3, 4]. As radiative recombination is enhanced at low relative energy, electron cooling must be stopped at an ion energy of 10 eV, before the high charge state fraction is excessively depleted. Electrons are separated or swept out of the trap and the so-called *resistive cooling* technique takes over, bringing the ion sample down to 4 K.

The Cooler Trap (see Fig. 1) is a cylindrical, axially-elongated structure immersed in a 6-T magnetic field provided by a superconducting, cryogen-free magnet. It has been calculated [2] that the 1.2-μs pulse can be efficiently captured with an effective trapping length of 400 mm when the energy of the incoming bunch has been reduced from 6 to 2 keV/u by lifting the bottom potential of the trap to 11 kV. Two endcaps of $L = 34.75$ mm and $d = 10$ mm are used to trap the bunch at the injection and act also as a differential pumping barrier. Simultaneous storage of particles of opposite sign is achieved thanks to the 21 inner electrodes, creating *nested traps* for both species (see Fig. 1). The inner electrodes have equal length $L = 17$ mm and diameter $d = 35$ mm and can create five nested traps (three for ions and two for electrons) with a fairly good harmonic potential in the center. This is advantageous since the axial oscillation frequency is then theoretically known. Moreover, due to the equal length of the electrodes, the same degree of harmonicity can be obtained at any axial position.

The storage time of at least 10 s (and possibly much more for long-term storage and on-demand delivery to the experiments) requires a high degree of vacuum not to lose the high charge state of the ion sample. Charge-exchange cross section calculations have shown that a vacuum better than 10^{-13} mbar is necessary [2]. Differential pumping of the volumes adjacent to the trap, as well as cryopumping in the trap environment, will be instrumental in matching the requirements.

The trap is at the moment in the assembly stage and the first tests will be run before the end of 2008.

RESISTIVE COOLING

Principle

Since no interaction of the considered species with other particles is involved, resistive cooling does not substantially alter the content of the ion sample and it is therefore preferable to the faster electron cooling at low energy. The energy loss occurs in an external circuit, where the image current arising from the ion motion is dissipated.

A charged particle in the vicinity of a conductor induces a charge density on the conductor itself. Since the potential on the conductor is defined, a theoretical solution to the apparent contradiction is the creation of an image charge of opposite sign which cancels the overall effect of the real charge on the conductor [5]. One can evaluate the induced image charge by writing the potential given by both the real and image particle and then imposing that the potential is zero on all points of the conductor. An analytical solution is possible for simple geometries, like that of a plate of radius R. Writing Gauss' law, the image charge q_{im} generated by a real charge q at a distance z is [6]

$$q_{im} = -\int_0^R \varepsilon_o E_z 2\pi r dr = -q\left(1 - \frac{z}{\sqrt{R^2+z^2}}\right) \tag{1}$$

with E_z normal component of the electric field with respect to the plate. In the case of a trap, the boundary is given by cylindrical electrodes and the solution is less trivial. The image can anyway be written as

$$q_{im,ring} = \gamma \frac{qR^2}{2} \int_{z_1-z}^{z_2-z} \frac{1}{(R^2+z'^2)^{3/2}} dz' = \gamma \frac{q}{2}\left[\frac{z'}{\sqrt{R^2+z'^2}}\right]_{z_1-z}^{z_2-z} \tag{2}$$

with z axial position of the particle with respect to the trap center and z_1, z_2 coordinates of the ring electrode. Here we considered a particle in the center of the ring, which is justified since by Gauss' law integration over 2π must be the same for a centered as well as for a radially displaced particle. The effect of the continuous ring of image charges outside the electrode is accounted for with the factor γ, which depends on the particular geometry and can be estimated numerically.

Connecting the pick-up electrode to an external circuit, the axial motion of a charge with instantaneous velocity v_z in the trap gives rise to an *image current* i_{im}

$$i_{im} = \frac{d}{dt}\Delta q_{im} = \frac{\partial}{\partial z}\Delta q_{im} \cdot \frac{\partial z}{\partial t} = v_z \frac{\partial}{\partial z}\Delta q_{im} \qquad (3)$$

which can be detected or dissipated on a resistance **R** [7]. The energy decay can be estimated in the approximation of a particle sinusoidally oscillating between two infinite plates at a distance D. Since the image current is then simply $i = qv_z/D$, the ohmic dissipation is $P = \mathbf{R}\langle i \rangle^2$ and $v_z^2 = 2E_z/m$ (with E_z axial component of the particle energy), the cooling time constant is

$$\tau = \frac{mD^2}{\mathbf{R}q^2}. \qquad (4)$$

If more particles are stored, the same argument holds for the center of mass (CM) motion of the cloud, which will be damped at the same rate [8], but in the linear approximation the other plasma modes will not be affected at all. This approximation is valid also for a real trap geometry if D is replaced by an effective equivalent distance and the oscillations are small, so that the induced current has still a linear dependence on v_z. In a real trap the nonlinear relation between particle velocity/position and image current, like in Eq. (2), anharmonicities in the electric field and the influence of the space charge of the HCI cloud on the overall electric field have to be taken into account. These nonlinear effects dissipate other components of the cloud motion beyond the CM oscillation. However, those higher-order contributions are in general far smaller, resulting in a long cooling tail.

An important remark concerns the external dissipating resistance. This is generally an RLC circuit in order to increase the peak resistance **R**. Therefore only the particles with axial frequency ω_z coinciding with the resonant ω_{RLC} of the circuit will experience the maximum cooling force. A second essential issue to remark is that the cooling being only in the axial direction, coupling of the axial oscillation to the motions in the radial plane will be needed to cool these degrees of freedom, too.

Simulation

Since the computation of the full dynamics of 10^5 particles is not feasible, we have used a Particle-In-Cell (PIC) code. A PIC code exploits the *mean field* approach to compute the dynamics of a system of charged particles moving in electric and magnetic fields without calculating all particle-particle interactions. It solves at each time step the Poisson equation on a spatial grid where the charge of the particles has been distributed, yielding the electrostatic potential including space charge effects. Particles are then advanced of one time step under the effect of the mean field. To reduce further the computation time, a limited number of *superparticles* is simulated: these carry the total charge of the real sample when the potential is evaluated but are advanced in time as normal particles.

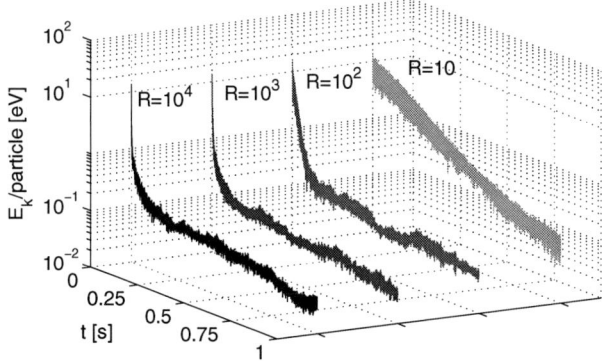

FIGURE 2. Mean kinetic energy of thirty C^{5+} ions during resistive cooling in a Penning trap by use of different ohmic resistances. Resistance values indicated in $M\Omega$.

The code used for our simulations [9] solves the Poisson equation in an rz-symmetric geometry with a Fast-Fourier-Transform (FFT) method [10]. The motion is nonetheless treated with a 3D algorithm. The choice of the algorithm must take into account the presence of a strong $\vec{B}$ field: an unacceptably small time step (inversely proportional to B) would be necessary with a conventional, non-adapted scheme. A Velocity Verlet (VV) algorithm adapted to this specific problem, whose accuracy is independent on the B strength [11], has been implemented in our code.

The resistive cooling routine is implemented in the code calculating at each time step i_{im} induced on the pick-up electrodes according to Eqs. (2) and (3) for all particles and then evaluating the voltage drop ΔV on the external circuit. ΔV is then applied back to the pick-ups, determining new boundary conditions for the solution of the Poisson equation. The effect is intuitive in a single-particle picture: an additional force pushing in the opposite direction of the ion motion.

The benchmark of the resistive-cooling PIC code considered the only published reference on resistive cooling in a Penning trap, i.e. the Mainz *g-factor* experiment on C^{5+} [12], where the cooling of 30 ions was observed. The full treatment will be the subject of a forthcoming article. We just show some results hinting at interesting features of this cooling technique. Due to accuracy loss and excessive computational effort, our simulations cannot cover the full time span of a minute observed experimentally. In order to have useful results, we generally used increased damping resistances **R** while maintaining all the other experimental parameters.

The single-particle cooling showed a cooling time constant $\tau = 1.34/\mathbf{R}$ s, with **R** in $M\Omega$, in perfect agreement with Eq. (4) and the experiment. More interesting is the cooling of 30 ions. Figure 2 shows the mean kinetic energy of 30 C^{5+} ions being cooled with different resistances. To simplify the problem, ohmic resistances in the range of $10 \div 10^4$ $M\Omega$ were used, i.e. the bandwidth of the cooling circuit was infinite. As expected from the theory, the reduction in **R** severely affects the initial phase of the process, where the CM motion is rapidly cooled. The time before the slow cooling tail sets in is increased with increasing **R** and the tail is not visible for **R** = 10 MΩ. But as the diagram indicates, the time constant of the tail τ_{tail} shows a very weak dependence on **R**: its values are 0.93, 0.84 and 0.65 s for 10^2, 10^3 and 10^4 MΩ, respectively. This qualitatively suggests that a broadband circuit overlapping with the frequency spread due to space charge may have beneficial effects despite a lower peak resistance.

CONCLUSIONS AND OUTLOOK

We have given an overview of the HITRAP project and facility, which will offer new experimental possibilities with HCI at low energy. The Cooler Trap was presented and the cooling processes were introduced. We have shown that with the help a a PIC code the resistive cooling technique can be investigated. Preliminary results hint at fruitful applications of the technique within HITRAP and further.

ACKNOWLEDGMENTS

G. Maero would like to thank G. Zwicknagel and R.C. Thompson for intensive and fruitful discussion.

REFERENCES

1. T. Beier et al., *Phys. Rep.* **339**, pp. 79–213 (2000).
2. F. Herfurth et al., "Highly Charged Ions at Rest: the HITRAP Project at GSI" in *Physics with ultra slow antiproton beams*, edited by Y. Yamazaki and M. Wada, AIP Conference Proceedings 793, American Institute of Physics, New York, 2005, pp. 278–290.
3. G. Zwicknagel, "Electron Cooling of Highly Charged Ions" in *Non-neutral Plasma Physics*, edited by M. Drewsen, U. Uggerhoj and H. Knudsen, AIP Conference Proceedings 862, American Institute of Physics, New York, 2006, pp. 281–291.
4. R. Nersisyan C. Toepffer, and G. Zwicknagel, *Interactions Between Charged Particles in a Magnetic Field*, Springer, Berlin-Heidelberg, 2007.
5. J.D. Jackson et al., *Classical Electrodynamics (second edition)*, John Wiley & Sons, New York, 1975.
6. B.I. Bleaney and B. Bleaney, *Electricity and Magnetism*, OUP, Oxford, 1989.
7. H.G. Dehmelt and F.L. Walls, *Phys. Rev. Lett.* **21**, pp. 127–131 (1968).
8. F.G. Major, V.N. Gheorghe, and G. Werth, *Charged Particle Traps*, Springer, New York, 2004.
9. G. Maero et al., "Simulation of Cooling Mechanisms of Highly-Charged Ions in the HITRAP Cooler Trap" in *Proceedings of COOL'07*, edited by R.W. Hasse and V.R.W. Schaa, JACoW, http://www.JACoW.org, 2008, pp. 130–133.
10. R.W. Hockney, *J. Ass. Comp. Mach.* **12**, pp. 95–113 (1965).
11. Q. Spreiter and M. Walter, *J. Comp. Phys.* **152**, pp. 102–119 (1999).
12. H. Häffner et al., *Eur. Phys J. D* **22**, pp. 163–182 (2003).

Investigation of Space-Charge Phenomena in Gas-Filled Penning Traps

Sven Sturm[a], Klaus Blaum[a,b], Martin Breitenfeldt[c], Pierre Delahaye[d], Alexander Herlert[d], Lutz Schweikhard[c], Fredrik Wenander[d]

[a]*Johannes Gutenberg-Universität Mainz, 55128 Mainz, Germany*
[b]*Max-Planck-Institut für Kernphysik, 69117 Heidelberg, Germany*
[c]*Ernst-Moritz-Arndt-Universität, 17486 Greifswald, Germany*
[d]*CERN, 1211 Geneva 23, Switzerland*

Abstract. The centering of ions in Penning traps by a quadrupolar radiofrequency excitation in the presence of a buffer gas has been studied in the regime of high charge-densities. It is found to deviate significantly from the single-particle situation. In particular, the efficiency of the cooling process is affected as well as the resolving power. The behavior has been studied experimentally at the preparation trap REXTRAP and the high-precision Penning trap setup ISOLTRAP both located at the on-line mass separator ISOLDE at CERN. In addition, the phenomenon has been investigated numerically by a custom-designed simulation.

Keywords: Penning trap, space charge, cyclotron cooling
PACS: 07.75.+h, 37.10.Ty, 85.30.Fg.

INTRODUCTION

The combination of azimuthal quadrupolar excitation with buffer-gas damping leads to a very effective centering (also called axialization) of ions in Penning traps [1]. The method has been developed at the ISOLTRAP mass spectrometer [2] in the context of the study of masses of short-lived-nuclei [3-5]. After its introduction it has been adopted [6] and found widespread application in the field of analytical (bio-)chemistry by Fourier-Transform Ion Cyclotron Resonance Mass Spectrometry (FT-ICR MS) [7].

Recently the technique came back to nuclear physics for the preparation of radioactive ion beams [8]. The present study considers the implications of space charge, i.e. of the presence of many ions of either the particles of interest or, in addition, contaminating species.

EXPERIMENTAL RESULTS

The Coulomb-force interaction between stored ions can alter their behavior with respect to that of the single-ion motion, resulting in shifts of the radial and axial eigenfrequencies. These shifts have been measured at the Penning trap systems REXTRAP [9] and ISOLTRAP [2] located at CERN. As an example, Fig. 1 shows the ion yield of $^{41}K^+$ ions stored in and extracted from ISOLTRAP's preparation Penning trap after the application of a quadrupolar rf excitation at the sum of the reduced cyclotron and magnetron frequency $v_+ + v_-$ [10]. In the right-hand side plot, the total amount of ions is four times higher than in the left-hand side picture. Ions of a natural mixture of both stable isotopes, ^{39}K and ^{41}K, were injected prior to the application of a dipolar rf excitation to increase their magnetron radius. With a subsequent mass-selective quadrupolar rf excitation, only the $^{41}K^+$ ions are supposed to be recentered. However, for a large amount of stored ions, i.e. a large space-charge due to the simultaneously stored $^{39}K^+$ ions, a shift and a splitting of the resonance frequency is observed. [11]

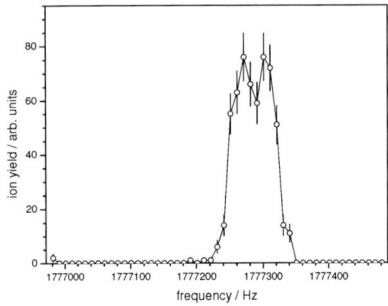 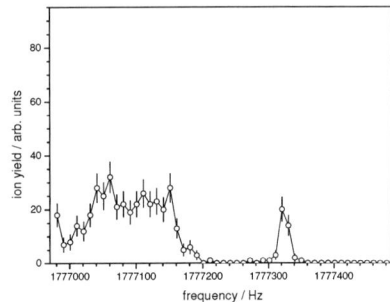

FIGURE 1. Ion yield of $^{41}K^+$ ions as a function of the frequency of a quadrupolar rf excitation. In the right-hand side graph, the total amount of initially stored ions is four times higher than in the left-hand side graph.

SIMULATIONS

The observed shifts of the resonance frequencies have been reproduced by numerical simulations. For medium-size ion clouds a specialized scheme has been developed which is capable of simulating the ion trajectories in Penning traps, taking into account a broad range of imperfections. The electrostatic potential can be included as numerical data extracted from an external Poisson-solver and the space-charge interaction is handled with an adjustable accuracy. The particle-number scaling is of the order of $N \log N$, allowing for up to 10^3 or 10^4 particles to be simulated directly, or about 10^7 particles by use of a rough Coulomb-scaling approach. In that regime a wide range of interesting effects can be addressed. The semi-direct force calculation allows the accurate handling of inter-particle interaction as, e.g., hard collisions including the resulting phase-decoherence, which is neglected in macroscopic approaches as the "particle in cell"-method (PIC). Furthermore no assumptions are made on the symmetry of the ion cloud in the trap, which is an important precondition for the simulation of externally excited particle clouds.

Numerical Particle Simulation via Time-Domain integration

The main difficulty in the simulation of ion trajectories in the presence of space-charge is the implementation of a scalable interaction calculation. The direct calculation of the Coulomb interaction between all particles leads to numerical complexities of the order of $N!$ or N^2 and thus to extreme computation times with particle numbers $N > 10^4$. An additional difficulty is introduced by the relative strengths of the Lorentz and Coulomb force ($F_L/F_C=(2\omega_c\omega_+)/\omega_z^2 \approx \omega_+/\omega_- \sim 10^3\text{-}10^4$), producing completely different time scales for the radial eigenfrequencies v_+ and v_-. For the numerical simulation this translates to a so-called stiff differential equation and thus to a large number of time-steps when trajectories for typical experimental durations (10 ms-1 s) are calculated. Since the effects of interest are usually observed in resonance spectra obtained by scanning the excitation frequency, the simulation has to be carried out numerous times in order to obtain a single spectrum. Thus, the calculation time scales linearly with the number of data-points in the spectra.

This trivial data-parallelism can be exploited by simply running the simulation parallel on different computers, each with a different excitation frequency. This kind of data-parallelism directly leads to speed-ups proportional to the number of available CPUs N_{CPU} up to typically $N_{CPU} = 41$ (the number of frequency steps in a typical ISOLTRAP spectrum), whereas the much more complex real task parallelism typically yields significantly lower speed-ups.

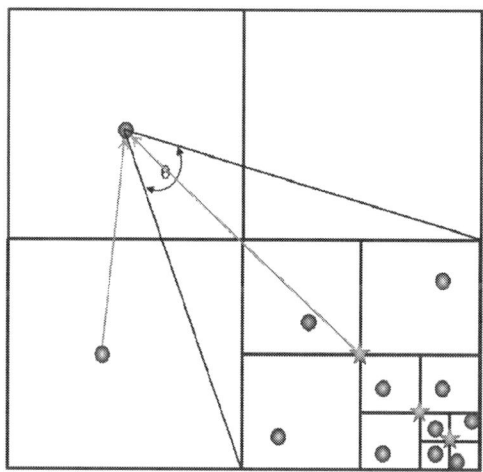

FIGURE 2. Division of the trap space into boxes. The stars symbolize the calculated multipoles.

Application of an Adapted Barnes-Hut Treecode Algorithm

The Barnes-Hut algorithm or more general the Fast Multipole Methods (FMM) allows to strike a balance between the macroscopic force calculation as used in PIC codes which is ignoring single-particle interaction and the microscopic direct Coulomb-force calculation which is too slow for full-scale simulations. It yields an approximation of the force at the position of each particle with an arbitrary maximum deviation and does typically scale with $N \log N$, depending on the momentary spatial particle distribution. The trap volume, which is without loss of generality divided into eight equally sized boxes. Those boxes containing more than one particle are further subdivided until each box contains exactly one particle or non (Fig. 2). In a bottom-to-top algorithm the first multipoles (typically mono- and dipole) of the lower-level boxes are calculated and assigned to the upper-level box, building a tree-structure. Since the multipoles can be summed easily the calculation cost for building the tree is relatively low. The force on the individual particles can now be determined from the multipoles of the next-neighbor boxes, if the opening angle θ (see Fig. 2) is smaller than an arbitrary value, meaning that the calculation error will be small. If θ is too large the force has to be obtained from the multipoles of the lower-level boxes.

In general the number of tree levels that have to be evaluated will be $\sim \log N$. For large ion numbers the time needed for building the tree can be almost neglected since it has to be built only once per timestep and can be used for all particles. In Fig. 2 it becomes obvious that the distribution of particles in the trap volume is strongly correlated with the speed-up obtained by this method. Since the amount of data stored in the tree structure is relatively low, the method is easily useable in task-level parallel computation, which is exploited in the simulation.

Time Domain Integration via a Krylov-Subspace Solver

As mentioned above, optimization of the integration method is crucial for the feasibility of the simulation. Several integration methods have been evaluated for this specific problem. As mentioned, the Ordinary Differential Equation System (ODES) is quite stiff as a result of the large difference in the time scales of the magnetron and the cyclotron motion, so solvers with possible preconditioning were used. On the other hand the system size is enormous, so direct Matrix/Jacobian based solvers are excluded as the Jacobian would be of size N^2. A Krylov solver with banded preconditioning and adaptive stepsize was found to have the best performance. The calculation of the residuals in the Krylov subspace decreases the dimensionality of simultaneously needed data, especially in the case of the (almost) block-diagonal Jacobian of the undisturbed Penning trap. Space-charge interaction introduces small off-diagonal elements but in most cases the banded preconditioner is still giving a remarkable speed increase.

For a fast and simple change of the solver the CVODE [12] package was used. The step size is automatically adopted by the solver on basis of a relative local tolerance, which was set to 10^{-8} for most simulations. The resulting numerical error was estimated by the amount of energy conservation violation in the undisturbed trap. Surprisingly, a standard Runge-Kutta 4th order algorithm with adaptive step size without preconditioning was just a few percent slower than the Krylov solver. In order to be able to use a 1st order ODE solver, the equations of motion for the Penning trap had to be transformed into a $6N$-dimensional vector of 1st order ODEs.

Coulomb Scaling

Even with a highly optimized code it is still not possible to simulate ion clouds in the typical size of $N \sim 10^4$-10^9 ions for the REXTRAP experiment. For that reason a linear Coulomb scaling is used in addition to the methods discussed above. The simulation is carried out with a number N of particles and the force on each particle is scaled by an arbitrary factor. In that way mean field effects should be reproduced rather accurately, whereas hard scattering and multi-particle interactions might be significantly altered.

Model-Based Mean-Field Simulation

In a different approach, the mean field of a large ion cloud can be simulated by fitting an assumed shape to the particle distribution. In the case of the quadrupolar recentering excitation in the presence of buffer-gas a well fitting shape is an ellipsoid elongated in the radial plane of the trap as shown in Fig. 3.

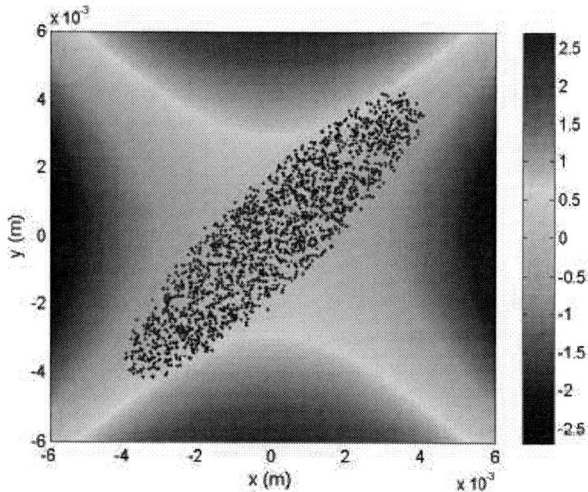

FIGURE 3. Simulated ion distribution after quadrupolar excitation near the resonance frequency v_c. The color illustrates the electric field strength of the quadrupolar excitation.

The mean field force can now be calculated from the analytical solution of the Poisson equation for the chosen shape. Since only a small number of ions have to be simulated in order to obtain a good fit, the speed-up with this method can be enormous. The resulting trajectories have to be checked for self-consistency. As a drawback, the direct ion-ion interaction gets lost by using this method.

SIMULATION RESULTS

When applying the quadrupolar recentering method in the presence of a large number of ions (of the same species), space-charge interaction causes the resonances to deviate significantly from the theoretical line shape of single events. The line-width increases drastically and the line becomes asymmetrically broadened towards higher frequencies which leads to a corresponding overall shift. Figure 4 shows the result of such a calculation (left) as compared to a measurement at REXTRAP (right) The simulation reproduces the observed effects. Similar good agreement has been found for the simulation of contaminent space-charge effects as shown in figure 1 [13].

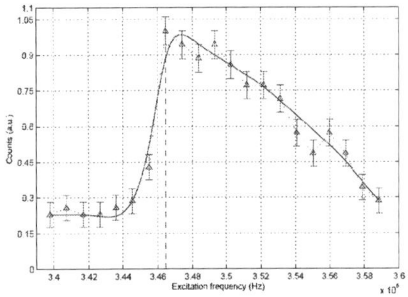 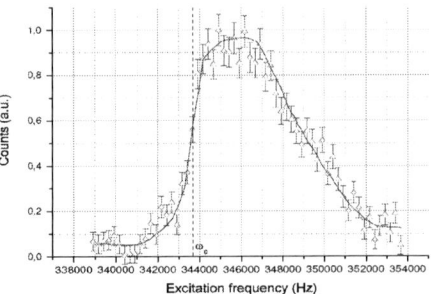

FIGURE 4. Simulated (left) and measured centering resonance with $\sim 10^7$ $^{133}\text{Cs}^+$ ions in REXTRAP (right). The resonance is broadened to higher frequencies and shifted from the expected single-particle resonance frequency $v_+ + v_-$. The magnetic field strength was not adjusted to the experimental conditions, hence leading to a difference in the absolute frequencies.

ACKNOWLEDGMENTS

This work was supported by the German Federal Ministry for Education and Research (BMBF) under contract no. 06GF186I and 06MZ215, by the Helmholtz Association of National Research Centers (HGF) under contract no. VH-NG-037, and by the European Commission under contract no. HPRI-CT-2001-50034. The authors acknowledge the support of the technical group of ISOLDE. S.Sturm acknowledges financial support by CERN.

REFERENCES

1. G. Savard et al., *Phys. Lett. A* **158**, 247 (1991).
2. M. Mukherjee et al., *Eur. Phys. J. A* **35**, 1 (2008).
3. K. Blaum, *Phys. Rep.* **425**, 1 (2006).
4. D. Lunney et al., *Rev. Mod. Phys.* **75**, 1021 (2003).
5. Special issue on Ultra-Accurate Mass Spectrometry and Related Topics, edited by L. Schweikhard and G. Bollen, Int. J. Mass Spectrom. 251, Elsevier, Amsterdam, 2006.
6. L. Schweikhard et al., *Int. J. Mass Spectrom. Ion Processes* **120**, 71 (1992).
7. S.H. Guan et al., Chem. Rev. 94, 2161 (1994); A.G. Marshall et al., Mass Spectrometry Rev. 17, 1 (1998); A.G. Marshall, Int. J. Mass Spectrometry 200, 331 (2000)
8. G. Bollen, S. Schwarz , Ion traps for radioactive beam manipulation and precision experiments, Nucl. Instr. Meth. B 204, 466 (2003)
9. D. Habs et al., *Nucl. Instrum and. Meth. B* **139**, 128 (1998).
10. G. Bollen et al., *J. Appl. Phys.* **68**, 4355 (1990)
11. L.Schweikhard et al., Eur. J. Mass Spectrom. 11, 457 (2005)
12. A. C. Hindmarsh et al. *SUNDIALS: Suite of Nonlinear and Differential/Algebraic Equation Solvers.* ACM Transactions on Mathematical Software 31(3), 363 (2005).
13. S. Sturm et al., publication in preparation.

Electrodynamics of neutron star magnetosphere: an example of non-neutral plasma in astrophysics

Jérôme Pétri

Centre d'étude des Environnements Terrestres et Planétaires, 10-12, avenue de l'Europe, 78140 Vélizy, FRANCE
Laboratoire de Radio-Astronomie, École Normale Supérieure, 75005 Paris, FRANCE

Abstract. Although discovered almost forty years ago, pulsars still rank amongst the most fascinating astrophysical objects in the universe. They are believed to be strongly magnetized rotating neutron stars, whose existence had already been predicted soon after the discovery of the neutron in the beginning of the 30s. However, relatively little progress has been made in understanding the fundamental physical mechanisms at work. Due to the lack of a satisfactory equation of state at very high densities, the internal structure of the neutron star remains enigmatic. Moreover, a global self-consistent picture of the close surrounding of the star, responsible for the emission of electromagnetic waves as well as for the launch of the pulsar wind and the energy loss due to interaction with the ambient medium, has not yet been proposed. Here, we give a brief review on the theory of pulsar's electrosphere (and magnetosphere) as well as on some recent developments. After a historical introduction, we recall some basic properties of these compact objects. We then present our understanding of the pulsar magnetosphere. Some promising recent developments will be discussed in detail: the so-called electrospheric model.

Keywords: Pulsars — Magnetosphere — Plasmas — Magnetohydrodynamics
PACS: 97.60.Gb; 095.30.Qd; 2.60.Cb; 52.27.Jt

INTRODUCTION

Discovery of the first pulsar

July 1967, the radio telescope at the radio astronomical observatory at Mullard in England is ready to acquire its first data. Formed by a matrix of 2048 dipolar antennas spreaded over more than 4.5 acres and operating at 81.5 MHz with a pass-band of 1 MHz, it was conceived to study the structure of radio compact sources by observing their scintillation caused by irregularities in the interplanetary medium. Indeed, radio waves emitted by a far-away source and propagating through the solar wind are subject to scintillation.

Jocelyn Bell, Ph.D. student at Cambridge observatory at this time, undertook a systematic search of such sources in November of the same year. She noticed day after day the appearance of a strange periodic signal emanating from a fixed position in the sky with galactic coordinates

$$\begin{cases} \alpha_{1950} &= 19\text{h } 19\text{mn} \\ \delta_{1950} &= 21°\ 58' \end{cases}$$

This signal consisted of a series of pulses with duration approximately 0.3 s occurring regularly every 1.337 s with an extraordinary regularity. This appearance period of the pulse was extremely stable, see Fig. 1. After correcting from the Doppler effect caused by the orbital motion of the Earth around the Sun, [1] estimated the period to be $P_0 = 1.3372795 \pm 0.0000020$ s. This value is slightly incorrect. Indeed, two independent observations at Parkes in Australia and at Arecibo in Puerto Rico gave a better estimate to be $P_0 = 1.33730109 \pm 0.00000007$ s. Hewish and his collaborators forgot to count one unit in the record of 64 000 pulses acquired during a two week campaign. Any kind of interpretation has then been proposed to find a terrestrial origin to this phenomenon, from the interference with a radar to the reflection of terrestrial signals from the moon, without any success. The absence of parallax showed clearly that the source was located well outside our solar system. The shortness of the pulses emitted at all frequencies being less than 16 ms, the size of this source could not exceed a few thousand of km (approximatly 4800 km).

Soon after, 3 other sources were identified with astonishingly similar properties. Given their small size, astronomers looked for an explanation in terms of radial oscillations of white dwarfs or neutron stars. The first option was quickly withdrawn because such short periods could not be due to pulsations of white dwarfs, which oscillate at much longer periods than those observed. On the other hand, radial oscillations of neutron stars are too short.

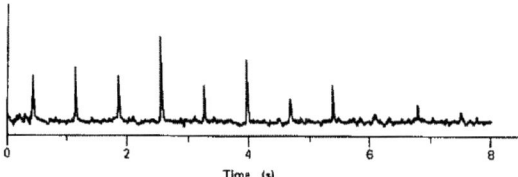

FIGURE 1. Radio signal measured from the first pulsar ever discovered, PSR1919+21, showing the regular arrival time of the pulse every 1.337 s.

The name *pulsar* given to these objects (an abbreviation of *pulsating radio source*) has its origin in this interpretation. This incorrect nomenclature is misleading because it is nowadays believed that a pulsar is a strongly magnetized neutron star possessing a high rotational speed. This rotation generates the pulses of the observed radio emission and has nothing to do with stellar pulsations as claimed before.

Pulsar's denomination

At the beginning, the name of a pulsar was formed by a combination between its approximated position in the sky preceded by the name of the observatory in which the pulsar was discovered. For example, the first pulsar was named CP 1919+21, CP being the initials for Cambridge Pulsar.

Nowadays, this terminology is standard, all pulsar names start with the prefix PSR (abbreviation for Pulsating Source of Radio) followed by 4 numbers indicating the right ascension α in coordinate 1950.0, followed by the declination δ and its value expressed with two numbers. Following these rules, the first pulsar was renamed by PSR 1919+21. If the resolution is not sufficient to distinguish between two of them, the declination is expressed in one tenth of degrees by adding a third number. When the coordinates are given with respect to the equinox of 1950.0, they are preceded by the letter B while if given with respect to the equinox of 2000.0, they are preceded by the letter J. The latter denomination is the one in use nowadays, by convention all pulsar names are given in coordinate 2000.0. Let's indicate for instance the standard names for the most famous and well studied pulsars which are

- the Crab : PSR 0531+21
- Vela : PSR 0833-45
- Geminga :PSR 0633-1746.

A simple definition

Our present knowledge on pulsar theory can be summarized very briefly in one sentence! A pulsar is a strongly magnetized rotating neutron star. It emits electromagnetic radiation in a broad frequency range from radio waves to X- and γ-rays. Neutron stars, just as white dwarfs and stellar mass black holes, are end-products of stellar evolution. They have unusual physical properties and are wonderful laboratories to study most extreme physical conditions, impossible to achieve under terrestrial conditions. Vacuum electric fields as high as 10^{12} V/m could develop at the surface of neutron stars. Fields of such strength can pull electric charges out of the neutron star's crust, even against the extraction work opposed by inter- and intra-ionic forces. As a result, charged particles freely expand in the star's close environment forming a space charge filled atmosphere which we refer to as an electrosphere. We now turn to a detailed discussion of such an electrosphere.

ELECTROSPHERE OF A PULSAR

A magnetosphere entirely filled with plasma is not the unique solution to the problem of the pulsar's magnetospheric structure. Indeed, in a previous work, [2] have shown that some solutions exist for which the magnetosphere is almost completely empty! They assumed an aligned rotator and derived a fully electromagnetic self-consistent model in which regions of non neutral plasma moving in differential rotation are separated by large vacuum gaps. Others also showed with help on numerical simulations that an entirely filled and corotating magnetosphere is electrodynamically unstable, see for instance [3]! In the following discussion, we make an explicit distinction between the magnetosphere and the electrosphere, the latter corresponding to regions of the former filled with non neutral plasma. In all aforementioned models, there was no need for a distinction between both because they were assumed to be perfectly identical. We undertook in deep details the study of the geometrical and dynamical properties of such an electrosphere. We first remind how to construct such a system by an iterative process. We then analyze the electromagnetic stability properties and demonstrate an important result for the physics of pulsars.

To build an electrosphere, we used the following procedure. Starting from a neutron star assumed to be a perfect conductor surrounded by vacuum, the rotation of the dipolar magnetic field induces a charge density on the stellar crust. The associated intense electric field then extracts the charge carrying particles whatever their nature (protons, electrons or ions). Particles follow the magnetic field lines until they reach an equilibrium state in which their electrostatic energy becomes minimal. Their density is estimated by the electric drift approximation (i.e. motion in the $\vec{E} \wedge \vec{B}$ direction, and, due to axisymmetry, motion in the azimuthal direction $\vec{e}_\varphi$) leading to a sharp boundary between space charge regions and vacuum (typical for non neutral plasmas, while for neutral plasmas the interface between charge and vacuum is always smooth). These charges exert a back reaction on the surface charge density which has then to adapt to the new space charge configuration. Then a new fraction of this modified stellar crust charge is extracted, the new equilibrium state with the associated charge density and boundaries has to be found, leading to a new differential rotation rate (deduced from the electric drift approximation) and so on. The iteration is stopped when all the charges on the star have disappeared. The only free parameter is the initial total charge of the neutron star (not necessarily set to zero).

The final result for a given total charge of the system is shown in Fig. 2. The star is entirely surrounded by charge-separated plasma consisting of an equatorial belt, positively charged and two negatively charged domes located over the polar caps (the south pole is not represented due to equatorial symmetry). It is shown that in order to cancel all the surface charge density, the star must be entirely surrounded by corotating electrospheric plasma. These two charged regions are separated by large vacuum gaps. Some inner part of the equatorial belt is directly connected along field lines to the star without any gap, and is thus corotating with it, while the outer part, which is not electrically connected by field lines to the star is super-rotating everywhere. An important result of this model is the differential rotation in the outer part of the disk. This is expected because [4] has shown that an electrosphere cannot entirely be in corotation with the star. The differential Goldreich-Julian density is sharply peaked at the innermost part of the differentially rotating region but never exceeds three or four times the corotation value. We have also plotted the residual stellar surface charge in comparison to the initial one in order to illustrate the quality of our numerical results: at the end of the iteration process no more than 1 % of the initial surface charge is left on the star's surface [5].

STABILITY ANALYSIS

Linear instability

We then studied in detail the consequences of such an electrospheric structure on the motion of charged particles in the star's environment. The differential over-rotation of the equatorial charged disk is of significant consequence on electric current flow in the electrosphere because it is subject to unstable motions, [6]. We have first conducted a detailed study of the global stability of this region by linear perturbation analysis.

The stability of a simplified model of the differentially rotating belt has been examined. The model consists of a charged thin plasma disc in differential rotation, embedded in a dipolar magnetic field anchored in the neutron star. In the linearized electrostatic perturbation approximation, this disc appears to be unstable to the so-called diocotron instability. The diocotron instability is an electrostatic instability induced by shearing flow in a non neutral plasma where particle inertia effects are neglected. It has a close analogy (at least in the linear analysis) with the more familiar Kelvin-Helmholtz instability met in fluid dynamics.

We found unstable modes to exist, the eigenfrequencies and growth rates of which have been determined. For

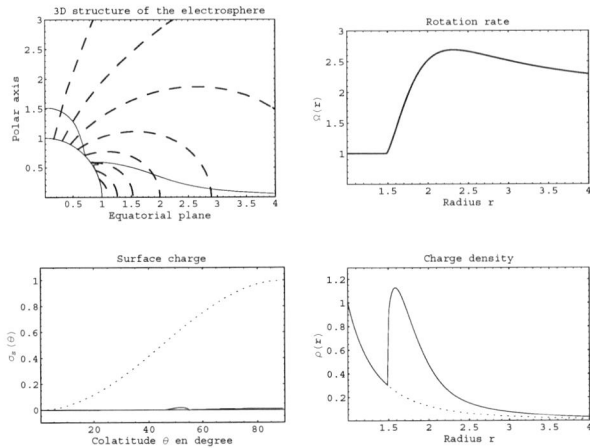

FIGURE 2. Electrospheric structure of an aligned pulsar without charge loss showing the geometry consisting of a dome and a disk (upper left picture), the differential rotation curve in the equatorial plane of the disk (upper right), the final (solid line) versus initial (dashed) crust charge density (lower left) and the differential (solid line) versus corotating (dashed) Goldreich-Julian charge density (lower right).

this purpose we have developed a numerical algorithm suited for the particular geometry of the electrosphere. The algorithm is simple, robust and accurate. The total charge of the star-electrosphere system appears to be the only relevant parameter. Stability has been discussed as a function of it: the smaller the total charge, the faster the growth of instabilities. When the total charge increases, growth rates decrease, but the instability nevertheless persists, [7, 8]. Increasing the total charge Q_{tot} has a stabilizing effect, in the sense that the growth rate of the fastest-growing eigenvalue decreases with total charge and eventually vanishes. When the system is unstable, the growth rates are of the order of the azimuthal rotation rate. This means that the diocotron instability, which develops in a few pulsar's periods, should be very efficient in causing particles to migrate across magnetic surfaces. Only rather large, coherent structures emerge. Indeed, the diocotron instability is due to collective behavior in the non neutral plasma implying relatively large scale structure [7].

Recently, we showed that relativistic and magnetic effects tend also to stabilize the diocotron regime, [9]. Close to the light cylinder, inertia of the particles becomes important as well as relativistic motion and magnetic field back-reaction. Thus we studied the most general instability known as the magnetron instability, [10]. This unstable structure can radiate electromagnetic waves to infinity. Nevertheless we found no appreciable difference in the growth rate between outgoing wave boundary conditions and outer conducting wall conditions.

Non linear evolution

The linear approach is insufficient to assess the long term behavior of the system. We therefore studied the long-time-scale evolution of this instability in the non-linear regime by means of both direct numerical simulations and a quasi-linear model. Non linear saturation of unstable modes and associated evolution of charge distribution in the disk have been studied by way of numerical simulations, which take full account of these non linearities. These simulations solve the electrostatics problem (which is elliptical in nature), coupled to the charge advection problem (which is hyperbolic in nature). We treated Poisson's equation by a Green's function method. To follow the charged particles motion, we used a conservative Godunov type scheme which solves the Riemann problem of charge conservation with second order corrections in space introduced by slopes limiters. To begin with, the equatorial disk has been evolved with no charge source. In these simulations, the charge distribution has been observed to evolve rapidly towards the formation of a few strong charge condensations which eventually orbite the star at high speed. The disk evolved into a few super-particles propagating like solitons! We then also considered the possibility that electric charge be continuously supplied to the disk by electron-positron pair avalanches, [11, 12]. These could be induced close to the neutron star by high energy photons traveling in the vacuum gaps. This other regime gives rise to an entirely different evolution. The instability then develops into relatively smaller scale electrostatic turbulence. The electric fluctuations induce a net equatorial charge flux, i.e. an electric current, leaving the pulsar's environment. The main effect of this instability is to make it possible for charged particles to diffuse cross-fields. The rate of diffusion remains however rather low and the associated diffusion coefficient relatively weak, [13].

We showed that, when the disk is externally fed with charged particles produced by a moderate pair creation activity in the magnetosphere, the diocotron instability causes diffusion of the charged particles across the magnetic field lines outwards. An equatorial cross-field electric current is observed to form, carrying a net charge flux radially outwards. This constitutes a hitherto ignored charge transport mechanism in the pulsar magnetosphere. We briefly discuss how this turbulent charge transport mechanism could bear on the problem of electric current closure in pulsar's magnetospheres. We have simulated a disk which suffers charge feeding from an external source with particles of the sign of its own surface charge, which in our simulations is positive. By this we mean to represent the effect of electron positron pair creation in the gaps existing between the domes and the equatorial belt [5]. Feeding the disk with charges by this process could maintain its structure unstable with respect to the diocotron instability at a low level, such that the lepton plasma production would give rise to a cross-field charge flux, even in the inner magnetosphere. Introducing a source of charges in this model is a simple first step towards the study of the interplay between pair creation and the diocotron instability.

The charge deposition on the disk is accompanied by the creation of an equal amount of charge of opposite sign. If it is assumed that the system suffers no net charge loss, this complementary charge would appear as surface charge on the stellar surface due to the electrostatic influence exerted by the disk. If, by contrast, it is assumed that this complementary charge, created simultaneously with the charge feeding the disk, is lost in some pulsar wind emitted from the polar caps, there would be a net charging of the star-electrosphere system, until the disk has grown so large that it reaches the light cylinder and the deposited charge is also lost at this outer boundary to the wind. In our simulations this eventual equilibrium regime has not been reached, although some substantial charge escape through the disk occurred on the long term.

We have studied the influence of the initial disk extension on the development of the diocotron instability. Several initial density profiles have been considered, along with different initial radial dimensions of the disk. To observe the eventual stationary state, in which significant loss of charges through the grid boundaries is to occur, the duration of the simulation would have to be increased by a factor of order 10 or more. This implied too long a computing time and too large data files. We therefore had to turn to a more efficient approach for the long term evolution of the system. We compromised by developing a quasi linear model as an alternative to the 2D non linear simulation. The quasi-linear model reduces the description of the system to just a few azimuthally averaged quantities. In the present geometry it is one-dimensional in space and therefore much less expensive in computer time than the non-linear 2D model. In a number of cases this model reproduces the results of the full non-linear 2D calculation very well, in the time interval of overlap.

Quasi-linear model

To better assess the consequences of the diocotron instability in this regime, and also to cut on the computing time of our full non linear simulation, we have developed a quasi-linear model of the development of this instability. Quasi-linear theory is justified when many azimuthal modes are excited. This has been checked by means of the fully non linear simulation. The quasi-linear approximation consists in following the azimuthally averaged part of the disk's charge distribution neglecting bilinear mode coupling terms in the evolution of non axisymmetric modes. Bilinear terms are however retained in the description of the evolution of the axisymmetric part. They give rise to effective diffusion, caused by the electrostatic non-axisymmetric perturbations. The calculation of the system's evolution in the quasi-linear framework boils down to an analysis of the rate of linear growth for the charge density profile, as it is at any given time. The azimuthally averaged charge density evolves due to diffusion in the associated electrostatic non-axisymmetric perturbations. The growth rates of these fluctuations are time dependent and tend to evolve to marginal stability. However, due to charge injection in the disk, marginal stability is never strictly reached and a stationary diffusive charge flux appears instead. We solved the quasi-linear evolution problem numerically and could thus evaluate the stationary value of the diffusion coefficient and the radial electric current which results asymptotically.

Obviously, the loss of charge through the equatorial disk cannot be permanent, since it implies a secular evolution of the total charge of the system. If this total charge were to grow large enough, positive charged particles from the disk would eventually be trapped by the system's large monopolar electric field. If such a state would be reached the pulsar's activity would cease. However, the system does not merely reduces to its equatorial disk. The equatorial loss of positive particles may, or should, be accompanied by the development of a negatively charged wind escaping from the poles, so that the current system is eventually globally closed. The electrosphere would then be in a stationary state, though still with large vacuum gaps under considerable potential drop, partially active to pair avalanching.

When only a few unstable modes are excited, the instability evolves to form coherent non linear structures in the charge density. When enough modes are excited and their phases are sufficiently incoherent, however, it is possible to describe the evolution in an approximate way in terms of a few azimuthally averaged macroscopic quantities only, such as the charge density and the electric potential. The quasi-linear model describes the evolution of these quantities. Such a reduction is justified when the correlation length of the electric field fluctuations is small as compared to the radius, or when their correlation time is small as compared to the characteristic evolution time of the density profile. In the present situation, the applicability of the quasi-linear model might be questioned in view of the fact that structures with a certain degree of spatial coherence are observed to form in our non linear simulations. However their characteristic "life time" appears to be short as compared to the characteristic time observed for the evolution of the average density structure. We checked a posteriori that in the region of time overlap the results of the fully two-dimensional simulation and of the quasi-linear model agree. Due to the instability, the particles are transported outwards, causing an expansion of the disk. The associated current increases continuously with time. Although all modes were initially excited, most of them progressively disappear and only a few survive when the state of marginal stability is reached, maintaining diffusion across the magnetic field lines. An interesting phenomenon appears after a while. The charge deposited on the disk begins to escape out in significant amounts and irreversibly. We find a situation in which the system, whilst being fed with charges, succeeds in generating, by the effect of the diocotron instability, a sustained flux of charges outwards. No stationary state has been reached, because, at the end of the simulation, charge is still accumulating over the disk. Since continued growth of the accumulated charge is impossible, we expect that at extremely long times a truly stationary state will be reached. The non linear simulation shows charge structures which, at any given time, appear to be spatially coherent, with well-defined charge condensations which retain their identity for a few rotational times. However, after a time of this order, these structure disappear and are replaced by similar ones, with unrelated phases. We believe that this poor coherence time, as opposed to the rather large spatial coherence, is the reason why a quasi-linear description meets with some success.

Therefore, the evolution of the system is entirely different in the presence of an external source of charges feeding the disk with charges of its own sign. This situation would be representative of neutron star magnetospheres with gaps which are partially pair-producing. These sources of charges would be located close to the region of contact of the differentially rotating belt with the corotating magnetosphere. The mean radial extension of the electrospheric disk first continuously increases due to charge diffusion induced by the diocotron instability. In the long term, a radial charge flux is organized and maintained, the charge diffusion coefficient keeping a sizeable value, due to continued charge deposition. The instability has the effect of redistributing the electric charge over the disk, which in return causes the growth rates of unstable modes to decrease from an initial value of the order of the star's rotation frequency. The system approaches marginal stability, but does not exactly reaches it, due to continued charge feeding. In this asymptotic situation, the growth rates become small, but remain just large enough to maintain diffusion of the charged particles across the magnetic field lines. For a sufficiently large duration of the simulation, a permanent outwardly oriented electric current has been observed to form in the equatorial plane, carrying away a large fraction of the charge delivered by the source to the disk. This type of current circulation in the closed magnetosphere, due to the presence of the diocotron instability, has been given but little attention before, although it could bear on the problem of current closure of the electric circuit of an active pulsar magnetosphere because it breaks the need for currents to be field-aligned in the closed magnetosphere. The construction of an electric current model for an active pulsar involving this turbulent transport mechanism in magnetically closed zones with gaps and connecting self-consistently to magnetically open wind regions across the light-cylinder is an ambitious task. Our present contribution establishes, in the framework of the adopted model, the relevance of cross field conduction by non-neutral plasma turbulent processes. We believe that this idea of a diocotron-turbulent charge transport in the pulsar's electrosphere, when extended to a 3D electrospheric structure including the presence of charged polar domes, will open the way to a new understanding of the electrodynamics of the pulsar's magnetosphere [13].

CONCLUSIONS

We can sum up our contribution by saying that we have reached a detailed understanding of the structure of the pulsar's electrosphere and of electric charge circulation in it. The associated high energy radiation can be calculated as well, for any given system's parameters. The model which emerges from our analysis is original in several respects. First, the consequences for high energy radiation by the pulsar of the existence of large vacuum gaps had never been discussed before self-consistently in the appropriate two-dimensional geometry. Second, the circulation of an equatorial current caused by the diffusion of charged particles in the diocotron-unstable electric field fluctuations is a concept that has been hitherto evocated but ill studied. Our work has established the reality and the relevance of this physics in the pulsar's context. Our analysis places the most active region for high energy gamma ray and lepton pair creation near the star, between pole and equator, but neither at the polar cap nor at the light cylinder. In this respect it introduces new ideas for pulsar modeling by changing the closed field region of the magnetosphere, previously regarded as inactive, into a potentially important actor of pulsar's activity. We summarize below the essential features of the results thus far

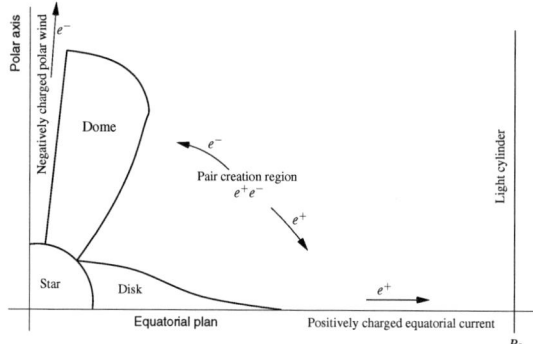

FIGURE 3. Electrospheric activity in the pulsar magnetosphere summarizing qualitatively the interaction of the different mechanisms believed to be at work.

obtained for the electrosphere of a pulsar and depicted schematically in Fig. 3

- large gaps separate the equatorial from the polar charge-separated plasma regions. The configuration of the equipotentials leaves no possibility for electric current to flow from domes to disc or conversely
- the electrosphere is finite in extent as long as the total charge of the system (neutron star+electrosphere) does not exceed three times the central point charge Q_c of the neutron star: $Q_{tot} \leq 3Q_c$. If moreover $Q_{tot} \ll 3Q_c$ the plasma is confined well within the light cylinder
- the part of the equatorial belt which is electrically connected to the stellar surface as well as the domes is in corotation. For $r \geq 1.5 R_*$ the belt is *super-rotating everywhere*, reaching a maximum of several times the stellar rotation velocity in the vicinity of $r = 2R_*$ and then decreasing slowly
- the maximum of the differential Goldreich-Julian density is in all cases the same. It never exceeds a few times the corotation Goldreich-Julian density.

In our opinion, it seems surprising to see that many important aspects of modeling a pulsar have been disregarded so far without justification. Many models deal only with the aligned rotator, assume axisymmetric flow, and a magnetosphere entirely filled with quasi neutral plasma, in the hope that the situation would not be very different from an oblique rotator. Should we emphasize that an aligned rotator is not a pulsar because it does not show any pulsed phenomenon! More likely, the opposite is the case. The nature of a pulsar is unavoidably connected with the physics of non-neutral plasmas, leading to a much richer description of the pulsar's magnetosphere (or better electrosphere). Large vacuum gaps exist in the magnetosphere, corotation with the star is not compulsory, actually differential rotation is inevitable in the presence of vacuum gaps. The diocotron instability, and its generalization, the magnetron instability, promise to give some new interesting results about the connection between the closed magnetosphere and the launching of the wind. It would be wrong to conclude that after almost 40 years of theoretical work on pulsar's physics, not much progress had been made towards a comprehensive description of this compact object.

REFERENCES

1. A. Hewish, S. J. Bell, J. D. Pilkington, P. F. Scott, and R. A. Collins. Observation of a Rapidly Pulsating Radio Source. *Nature*, 217:709, 1968.
2. J. Krause-Polstorff and F. C. Michel. Electrosphere of an aligned magnetized neutron star. *MNRAS*, 213:43P–49P, March 1985.
3. I. A. Smith, F. C. Michel, and P. D. Thacker. Numerical simulations of aligned neutron star magnetospheres. *MNRAS*, 322:209–217, April 2001.
4. W. G. Pilipp. On the Electrodynamic Equilibrium of a Space Charge Region around a Rotating Neutron Star with an Aligned Magnetic Field. *ApJ*, 190:391–402, June 1974.
5. J. Pétri, J. Heyvaerts, and S. Bonazzola. Global static electrospheres of charged pulsars. *A&A*, 384:414–432, March 2002.
6. R. C. Davidson. *Physics of non neutral plasmas*. Addison-Wesley Publishing Company, 1990.
7. J. Pétri, J. Heyvaerts, and S. Bonazzola. Diocotron instability in pulsar electrospheres. I. Linear analysis. *A&A*, 387:520–530, May 2002.
8. J. Pétri. The diocotron instability in a pulsar cylindrical electrosphere. *A&A*, 464:135–142, March 2007.
9. J. Pétri. Relativistic stabilisation of the diocotron instability in a pulsar "cylindrical" electrosphere. *A&A*, 469:843–855, July 2007.
10. J. Pétri. The magnetron instability in a pulsar's cylindrical electrosphere. *A&A*, 478:31–41, January 2008.
11. M. A. Ruderman and P. G. Sutherland. Theory of pulsars - Polar caps, sparks, and coherent microwave radiation. *ApJ*, 196:51–72, February 1975.
12. K. S. Cheng, C. Ho, and M. Ruderman. Energetic radiation from rapidly spinning pulsars. I - Outer magnetosphere gaps. II - VELA and Crab. *ApJ*, 300:500–539, January 1986.
13. J. Pétri, J. Heyvaerts, and S. Bonazzola. Cross-field charge transport by the diocotron instability in pulsar magnetospheres with gaps. *A&A*, 411:203–213, November 2003.

Next Generation Trap for Positron Storage

J. R. Danielson, T. R. Weber, and C. M. Surko

Department of Physics, University of California, San Diego, La Jolla, CA 92093

Abstract. Progress toward the development of a novel multicell Penning-Malmberg trap is described that will be capable of accumulating orders of magnitude more positrons than is possible presently. This design represents the next major step in antimatter storage technology. Experiments with test electron plasmas establishing techniques critical to the implementation of a practical multicell trap are presented. The latest design for a 21 cell trap capable of accumulating and storing more than 5×10^{11} positrons is described. This trap could facilitate multiplexing the output of the new generation of positron sources either operating now or currently under development, as well as the potential to provide record-high bursts of positrons for a variety of applications.

Keywords: Nonneutral plasmas, Penning-Malmberg traps, antimatter plasmas
PACS: 52.27.Jt, 52.25.Xz, 52.25.Fi, 52.25Kn

I. INTRODUCTION

The accumulation of low-energy positrons (i.e., the antiparticles of electrons) has become increasingly important in many fields, including atomic physics, plasma physics, and materials science. In the laboratory, positrons are now being used, or are planned for use, in many applications, including atomic and molecular physics [1,2], the formation of low energy antihydrogen [3,4], and the characterization of materials [5]. On the technological side, a new generation of more intense positron sources is under development [6-10]. Full utilization of these facilities will require the development of new techniques to accumulate and store large quantities of positrons.

Much progress in science with low-energy antimatter has been, and is expected to continue to be enabled by the development of novel nonneutral plasma techniques to create and manipulate single-component plasmas (SCP) in Penning-Malmberg traps, such as the one shown in Fig. 1. This device uses a uniform axial magnetic field and cylindrical electrodes with electrostatic potentials on the ends to confine particles of a single sign of charge (e.g., positrons, electrons, ions, or antiprotons) [1,11]. While the principal objective is the accumulation, manipulation and storage of positrons, the experiments described here use conventional single-component electron plasmas for increased data rate and ease of handling. In an actual positron application, the electron source would be replaced by bursts of positrons that are now accumulated routinely and efficiently using the buffer-gas trapping technique [12].

Here, we review recent progress toward what we regard as the next major step in antimatter storage technology, namely the development of a novel multicell trap that will be capable of accumulating orders of magnitude more positrons than is possible presently. This trapping concept was described in Ref. [13], and much of the more recent work reviewed here is described in more detail in Ref. [14]. When constructed, this multicell trap will provide record-high bursts of positrons for a variety of applications. It will also facilitate multiplexing the output of the new generation of positron sources either operating now or under development [6-10], thus enhancing their capabilities. The benchmark design goal described here is the trapping and storage of the order of 10^{12} positrons. They will be able to be stored for days and delivered rapidly for a variety of applications. The significance of this objective is illustrated in Fig. 2. Of considerable significance is the fact that the design is modular, and so larger or smaller antimatter storage capacities can be achieved by building the appropriate number of storage modules.

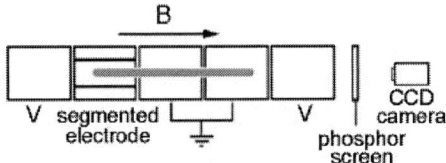

Figure 1. Schematic diagram of a Penning-Malmberg trap. As shown, it consists of five cylindrical electrodes, including one segmented azimuthally for radial plasma compression. Also shown is a phosphor screen and CCD camera that can be used to image radial density profiles.

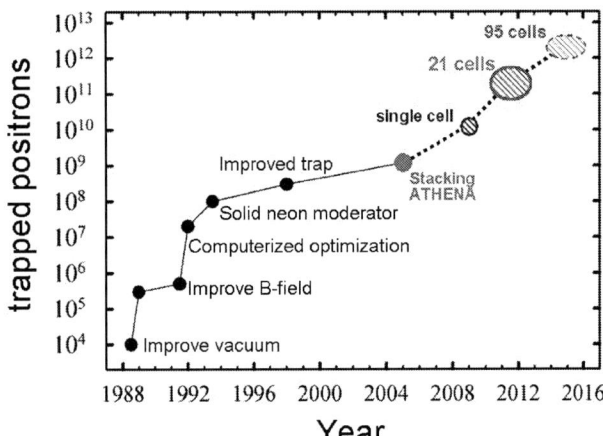

Figure 2. Progress in positron trapping from similar strength sources using a buffer-gas accumulator: actual record results (solid line), and projected capabilities (dashed lines). The projected capabilities of the 21-cell, multicell trap proposed here and a 95-cell trap (from Ref. [14]) are shown as shaded ovals.

II. MULTICELL TRAP CONCEPT

The concept of the multicell Penning-Malmberg trap is shown schematically in Fig. 3 [13,14]. The most severe limitation to confining plasmas consisting of large numbers of particles is that they have a large space charge potential, which in turn requires unacceptably large confinement voltages. In particular, for fixed plasma length, L_p, the number of particles, N, that can be stored in a trap is limited by the maximum potential, V_C, that can be applied to the electrodes. For a long, uniform plasma of radius R_p in a cylindrical electrode structure of radius R_w, the on-axis space charge potential, ϕ_o, (in Volts) is

$$\phi_o = 1.4 \times 10^{-7}(N/L_p)[1 + 2\ln(R_w/R_p)], \qquad (1)$$

where L_p is in cm. From Eq. 1, for example, for a plasma of 10^{10} positrons with $L_p = 10$ cm and $R_w/R_p = 10$, $\phi_o = 785$ V, which in turn, requires a value of $V_C \sim 800$ V. In a similar size trap, 10^{12} positrons would require $V_C \sim 80$ kV, which is impractical.

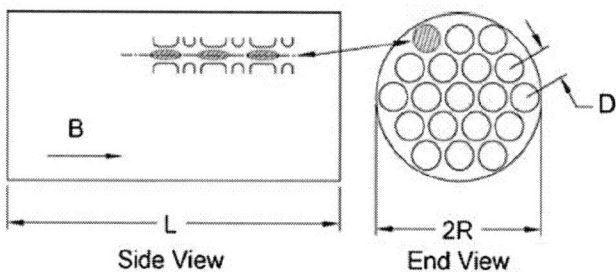

Side View   End View

Figure 3. Conceptual design of the multicell trap from Ref. [14], showing the arrangement of cells parallel and perpendicular to B. This early design had 19 hexagonally close-packed (HCP) cells perpendicular to B. The parameters for the 21 cell trap (i.e., 3 banks of 7 HCP cells) described here are summarized in Table I.

Table I. Design parameters of a 21-cell multicell trap.

Number of cells ($m \times p = 7 \times 3$)	21
Total positron number, N (10^{11})	≥ 5.0
Magnetic field (T)	5
Total electrode length, L (cm)	100
Electrode-package diameter, $2R$ (cm)	> 7.5
Plasma radius, R_P (cm)	0.2
Plasma length, L_p (cm)	20
Confinement voltage, V_c (kV)	1.0
Cell spacing (cm), D	2.0
Space charge potential (V)	750
Rotating wall frequency (MHz)	4

In the multicell trap, this space charge potential is mitigated by dividing the plasma into m, rod-shaped plasmas of length L, each oriented along the magnetic field [e.g., in a hexagonal-close-packed (HCP) arrangement transverse to the field as shown in Fig. 3]. These rod-shaped plasmas are shielded from each other by close-fitting copper electrodes. For a given maximum confining electrical potential, V_C, applied to the electrodes, the number of stored positrons will be increased by a factor of m. The multicell design also breaks up each long rod of plasma into p separate plasmas in the direction along the magnetic field (i.e., separated by electrodes at potential V_c). This reduces the effects of electrostatic and magnetic nonuniformities and the associated asymmetry-induced radial transport that limits the plasma lifetime.

We have proposed previously several arrangements including a 95 cell trap to store $\geq 1 \times 10^{12}$ positrons. Based upon recent experiments, we propose here an improved and considerably simplified design for a 21-cell multicell trap (m = 7; p = 3) that will be able to store almost as many particles (i.e., in excess of 5×10^{11}), as summarized in Table I. The principal design change comes from the recognition that a larger number of positrons can be confined in a single cell than was assumed previously, and this reduces the number of cells required for a given total accumulation capacity.

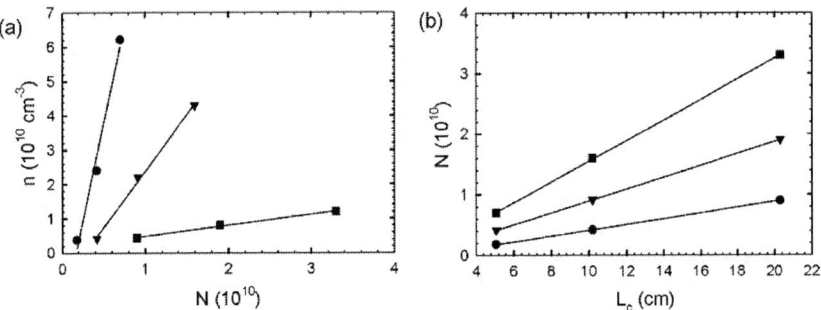

Figure 4. (a) Measured plasma density vs. total number, N, for three different confinement lengths L_c of (●) 5.1, (▼) 10.2, and (■) 20.3 cm; and (b) the dependence of N on L_c for three different fill voltages, V_f: (●) 300, (▼) 600, and (■) 900 volts. For all experiments, the confinement voltage $Vc = 1.0$ kV.

III. EXPERIMENTAL TESTS

There are a number of techniques that are necessary in order to realize a practical multicell trap. Here we describe several experiments that were done recently using test electron plasmas to demonstrate these techniques [14]. Specifically, we have explored the maximum trapping in a single cell, the operation of two (axial) plasma cells, off-axis plasma control, and plasma compression using rotating electric fields.

A. Trapping in a Single Cell

Plasmas were created and confined using a 1.0 kV confinement potential for three different confinement lengths, L_c = 5.1, 10.2 and 20.3 cm (see Ref. [14] for more details). The dependence of plasma density on the total number of particles, N, is illustrated in Fig. 4(a) for the three different confinement lengths, L_c. The dependence of N on L_c for the three different filling voltages, V_f, is illustrated in Fig. 4(b). Several of the cases exceed the $N = 1 \times 10^{10}$ particle design goal. These experiments have verified the operation of a single cell using parameters that are in the range of those necessary for a practical multicell trap.

B. Trapping in Two Storage Cells

The ability to create and manipulate two, in-line plasmas was also investigated [14]. Plasmas were loaded in two separate axial cells of the high-field trap using a "fill-shuttle-fill-hold" protocol. This experiment is illustrated in Fig. 5. Following the "hold" stage, independent control of the plasmas was demonstrated by depositing them sequentially onto the phosphor screen where they could be imaged separately. Each plasma consisted of $N \approx 5.5 \times 10^8$ electrons. The plasmas were 10 cm long, 0.93 mm in radius, and were separated axially by 10 cm. It was confirmed that there was no increase in outward radial transport with the addition of the second plasma to the trap. The plasmas were confined for 30 seconds with no noticeable expansion. Additional studies will be required to investigate confinement of these multi-cell, trapped plasmas at higher densities and on longer time scales.

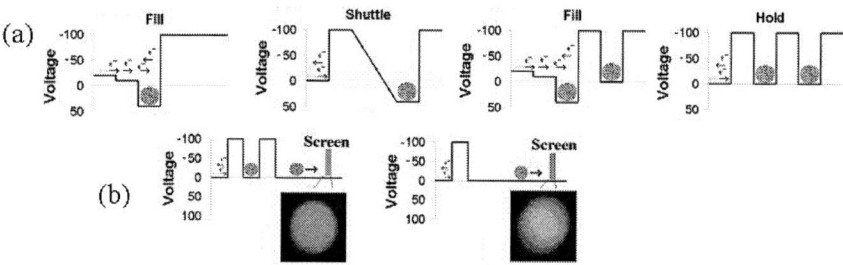

Figure 5. Schematic diagram of (a) the potential profile during the sequential filling of two, in-line plasma cells with electron plasma; and (b) the sequential dumping of these plasmas and the resulting images. From Ref [14]; see this reference for details.

C. Filling and Retrieving Plasmas from Off-axis Cells

One of the key features of the multicell trap is the utilization of off-axis cells for plasma storage. One critical aspect of this design is the ability to inject plasmas into the off-axis cells in an efficient manor and to be able to reverse this procedure to dump the contents of the cells. Recently we developed a method to transfer plasmas into and out of off-axis cells by the excitation and manipulation of a "diocotron" plasma mode [14]. This mode is the azimuthal rotation of a radially displaced plasma column around the center of the cylindrical electrode, excited by applying a low-frequency (typically ~ kHz) sinusoidal signal to a segmented electrode. The mode frequency is a nonlinear, increasing function of the plasma displacement, D, from the center of the electrode. When the drive frequency is swept from low to high, the mode can be made "autoresonant" with the drive signal so that the plasma locks to the drive [15,16]. The drive frequency determines the off-axis displacement, while the drive-signal phase sets the azimuthal position of the plasma. As illustrated in Fig. 6, plasmas can be translated radially across the field to a predetermined position, then dumped into an off-axis cell. Using this technique in reverse, plasmas can be extracted from off-axis storage cells, then returned to the symmetry axis for use at the output of the trap.

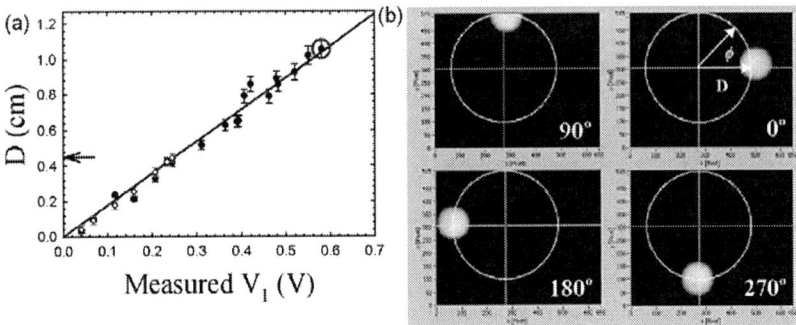

Figure 6. (a) Plasma displacement, D, as a function of the measured amplitude of the diocotron-mode signal. The circled point corresponds to 80% of the wall radius. (b) Plasma images for different drive-signal phases at a fixed displacement. From Ref [14]; see this reference for details.

D. Radial Plasma Compression with Rotating Electric Fields

In a Penning-Malmberg, the lifetime of the plasma is limited by the radial transport due to trap imperfections [17,18]. To mitigate this, we utilize the technique of rotating electric field compression (the so-called "rotating-wall" technique) [19-24]. This technique is now widely used and is cable of achieving essentially "infinite" confinement times. It relies upon an electric field, rotating in the plane perpendicular to the magnetic field, to inject angular momentum into the plasma and compress it radially. As originally developed [20,21], this technique required careful tuning to a mode in the plasma to compress weakly coupled plasmas (i.e., plasmas without crystalline ordering) that are relevant here. Recently, we discovered a new "strong drive" regime of operation that does not require mode tuning [22-24]. It works over a broad range of frequencies and compresses the plasma until the E x B rotation frequency, f_E, equals the applied rotating wall frequency, f_{RW}. The final density can thus be controlled by changing f_{RW} and does not depend critically on the RW amplitude. Furthermore, plasmas with a remarkably broad range of initial densities (e.g., varying by a factor of 20 or more) can be compressed by the application of a single, fixed value of f_{RW}.

This new regime of RW operation is expected to lead to considerable simplifications in the design of a practical multicell trap in that active control and interrogation of individual plasma cells is unnecessary. Very recently, much progress has also been made in understanding the coupling of the RW fields to the plasma [24]. This will aid greatly in being able to make accurate designs for advanced traps for long-term positron storage.

IV. NEXT GENERATION POSITRON TRAP

As indicated in Table 1 and shown schematically in Fig. 7, a 21-cell multicell trap will consist of 3 banks of storage cells with 7 individual cells in an HCP arrangement in each bank. The results shown in Fig. 4 indicate that with a 20 cm confinement length, we can store about 3×10^{10} particles per cell. With 21 cells total, this would result in the storage of a total of 6×10^{11} positrons. To fill this trap, we need a set of feed electrodes with a large radius to use the diocotron excitation technique to move the plasma off-axis and to perform the phase locked injection as shown in Fig. 6. After the first cell is filled, the "shuttle" procedure shown in Fig. 5 would be used to move the plasma from one bank of storage cells to the next. Finally, each individual cell would necessarily include an azimuthally segmented electrode for rotating electric field compression. This will allow for almost infinite confinement of the plasmas in each individual storage cell.

As shown in Fig. 7, each individual cell consists of four electrodes, including one segmented (multicolor) for the application of the necessary rotating electric fields. If more storage were desired, a larger bank of HCP cells could be used (e.g., the 19 cell arrangement in Fig. 3), or extra banks of cells could be added.

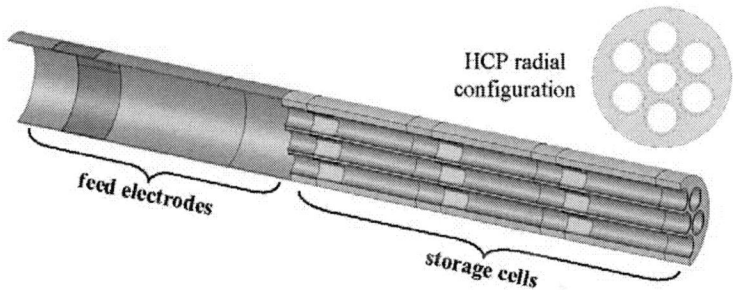

Figure 7. Schematic diagram of the 21-cell multicell positron trap, showing three banks of 7 cells in an hexagonally closed packed arrangement. Plasmas from the source will first enter the feed electrodes, then be moved off axis using autoresonant excitation of the diocotron mode to fill off-axis storage cells.

V. SUMMARY

It is our view that the development of the multicell Penning-Malmberg trap can represent the next major step in antimatter storage technology. The design described here would allow for the accumulation and storage of more than 5×10^{11} positrons. Experiments with test electron plasmas established several techniques critical to the implementation of a practical multicell trap. These experiments included storage of $>10^{10}$ particles in a single cell, operation of two cells simultaneously, and off-axis plasma manipulation for phase-locked plasma control.

One immediate application of this trap is to multiplex the outputs of the new generation of high-flux positron sources either in operation now [6, 7] or currently under development [8-10]. Further, the availability of such large numbers of positrons opens up many new possibilities, such as providing bursts of positrons far larger than available by any other means. Applications of these large pulses of positrons include the possible production of Bose-condensed positronium and the study of electron-positron plasmas. Lastly, the successful development of such a multicell trap will also be a major step toward the creation of a versatile *portable* antimatter trap.

ACKNOWLEDGEMENTS

We thank R. G. Greaves for helpful discussions and E. A. Jerzewski for expert technical assistance. This work was supported by NSF grants PHY 03-54653 and PHY 07-13958.

REFERENCES

1. R. G. Greaves and C. M. Surko, *Phys. Plasmas* **11**, 2333 (2004).
2. C. M. Surko, G. F. Gribakin, and S. J. Buckman, *Journal of Physics B* **38**, R57 (2005).
3. M. Amoretti, C. Amsler, G. Bonomi, *et al.*, *Nature* **419**, 456 (2002).
4. G. Gabrielse, N. Bowden, P. Oxley, *et al.*, *Phys. Rev. Lett.* **89**, 213401 (2002).
5. P. J. Schultz and K. G. Lynn, *Rev. Mod. Phys.* **60**, 701 (1988).
6. A. G. Hataway, M. Skalskey, W. E. Frieze, R. S. Valley, D. W. Gidley, A. I. Hawari, and J. Xu, *Nucl. Instrum. and Meth. in Phys. Res. A* **579**, 538 (2007).
7. C. Hugenschmidt, K. Schreckenbach, M. Stadlbauer, and B. Straßer, *Appl. Surf. Sci.* **252**, 3098, (2006).
8. A. P. Mills, Jr., D. B. Cassidy, and R. G. Greaves, *Mat. Sci. Forum* **445-446**, 424 (2004).
9. A. W. Hunt, L. Pilant, D. B. Cassidy, and K. G. Lynn, *Appl. Surf. Sci.* **194**, 296 (2002).
10. S. D. Chemerisov, C. D. Jonah, and H. Chen, *Nucl. Instrum. and Meth. in Phys. Res. B* **261**, 904 (2007).
11. C. F. Driscoll and J. H. Malmberg, *Phys. Rev. Lett.* **50**, 167 (1983).
12. T. J. Murphy and C. M. Surko, *Phys. Rev. A* **46**, 5696(1990).
13. C. M. Surko and R. G. Greaves, *Rad. Chem. and Phys.* **68**, 419 (2003).
14. J. R. Danielson, T. R. Weber, and C. M. Surko, *Phys. Plasmas* **13**, 123502 (2006).
15. J. Fajans, E. Gilson, and L. Friedland, *Phys. Rev. Lett.* **82**, 4444 (1999).
16. J. Fajans, E. Gilson, and L. Friedland, *Phys. Plasmas* **6**, 4497 (1999).
17. C. F. Driscoll, K. S. Fine, and J. H. Malmberg, *Phys. Fluids* **29**, 2015 (1986).
18. T. M. O'Neil, *Phys. Fluids* **26**, 2128 (1983).
19. X.-P. Huang, F. Anderegg, E.M. Hollmann, C.F. Driscoll, and T.M. O'Neil, *Phys. Rev. Lett.* **78**, 875 (1997).
20. F. Anderegg, E. M. Hollmann, and C. F. Driscoll, *Phys. Rev. Lett.* **81**, 4875 (1998).
21. E. M. Hollmann, F. Anderegg, and C. F. Driscoll, *Phys. Plasmas.* **7**, 2776 (2000).
22. J. R. Danielson and C. M. Surko, *Phys. Rev. Lett.* **94**, 035001 (2005).
23. J. R. Danielson and C. M. Surko, *Phys. Plasmas* **13**, 123502 (2006).
24. J. R. Danielson, C. M. Surko, and T. M. O'Neil, *Phys. Rev. Lett.* **99**, 135005 (2007).

List of Participants

Anderegg, Francois
Univ. of California San Diego
fanderegg@ucsd.edu

Anderson, Michael
Univ. of California San Diego
seriouslarry@hotmail.com

Apolinario, Sergio
Universiteit Antwerpen
sergio.apolinario@ua.ac.be

Aramaki, Mitsutoshi
Nagoya University
aramaki@nuee.nagoya-u.ac.jp

Bettega, Giovanni
Università degli Studi di Milano
giovanni.bettega@mi.infn.it

Bollinger, John
NIST, Boulder
john.bollinger@boulder.nist.gov

Brenner, Paul
Columbia University
pwb2103@columbia.edu

Castro, Jose
Rice University
Jose.Castro@rice.edu

Danielson, James
Univ. of California San Diego
jdan@physics.ucsd.edu

Dietrich, Matt
Univ. of Washington
dietricm@u.washington.edu

Drewsen, Michael
Univ. of Aarhus
drewsen@phys.au.dk

Driscoll, C. Fred
Univ. of California San Diego
cdriscoll@ucsd.edu

Dubin, Dan
Univ. of California San Diego
ddubin@ucsd.edu

Durand de Gevigney, Benoit
Columbia University
bd2194@columbia.edu

Eggleston, Dennis
Occidental College
dleggles@oxy.edu

Ennever, Paul
Columbia University
pce2101@columbia.edu

Fajans, Joel
Univ. of California Berkeley
jfajans@berkeley.edu

Gilson, Erik
Princeton Plasma Physics Lab.
egilson@pppl.gov

Hahn, Michael
Columbia University
mh2451@columbia.edu

Heidemann, Ralf
Max-Planck Inst. für Extraterrestriche Physik
Heidemann@mpe.mpg.de

Himura, Haruhiko
Kyoto Inst. of Tech.
himura@kit.ac.jp

Kabantsev, Andrey
Univ. of California San Diego
aakpla@physics.ucsd.edu

Kawai, Yosuke
Kyoto University
kwaiyosuke@h01a0395.mbox.media.kyoto-u.ac.jp

Killian, Thomas
Rice University
killian@rice.edu

Kurcz, Andreas
University of Leeds
A.Kurcz06@leeds.ac.uk

Kuroda, Naofumi
Inst. of Physics, Univ. of Tokyo
kuroda@radphys4.c.u-tokyo.ac.jp

LeSage, David
Harvard University
lesage@physics.harvard.edu

Maero, Giancarlo
GSI, Darmstadt
G.Maero@gsi.de

Marksteiner, Quinn
Columbia University
qrm1@columbia.edu

Marler, Joan
Aarhus University
joanmarler@gmail.com

Nelissen, Kwinten
Universiteit Antwerpen
kwinten.nelissen@ua.ac.be

Nikolaev, Eugene
Inst. for Energy Problems of Chemical Physics, Moscow
ennikolaev@rambler.ru

O'Neil, Tom
Univ. of California San Diego
toneil@ucsd.edu

Pedersen, Thomas
Columbia University
tsp22@columbia.edu

Pétri, Jérôme
Centre d'etude des Environnements Terrestre et Planetaires
Jerome.Petri@cetp.ipsl.fr

Pohl, Thomas
Harvard University
tpohl@cfa.harvard.edu

Porras, Diego
Max-Planck Inst. für Quantum Optics
diego.porras@mpq.mpg.de

Raithel, Georg
Univ. of Michigan
graithel@umich.edu

Roberson, Chuck
Office of Naval Research
croberson2@cox.net

Rolston, Steve
Univ. of Maryland
rolston@umd.edu

Romé, Massimiliano
University degli Studi Milano
Massimiliano.Rome@mi.infn.it

Rubin-Zuzic, Milenko
Max-Planck Inst. für Extraterrestriche Physik
mrz@mpe.mpg.de

Saitoh, Haruhiko
Atomic Physics Lab., RIKEN
saito@ppl.k.u-tokyo.ac.jp

Schweikhard, Lutz
Greifswald University
lschweik@physik.uni-greifswald.de

Stoneking, Matthew
Lawrence University
matthew.r.stoneking@lawrence.edu

Sturm, Sven
Johannes Gutenberg-Universitat Mainz
sturms@uni-mainz.de

Surko, Clifford
Univ. of California San Diego
csurko@ucsd.edu

Sütterlin, Robert
Max-Planck Inst. für Extraterrestriche Physik
robert@mpe.mpg.de

Vrinceanu, Daniel
Los Alamos National Laboratory
vrinceanu@lanl.gov

Weber, Toby
Univ. of California San Diego
tobyweber@gmail.com

Wurtele, Jonathan
Univ. of California Berkeley
wurtele@socrates.berkeley.edu

Yeliseyev, Yuriy
Kharkov Inst. of Physics and Tech.
eliseev2004@rambler.ru

AUTHOR INDEX

A

Anderegg, F., 89
Anderson, M. W., 114
Aramaki., M., 19
Avril, A., 25

B

Barna, D., 157
Beige, A., 31
Berkery, J. W., 55, 63, 75, 81
Bettega, G., 96
Blaum, K., 185
Blinov, B. B., 25
Boozer, A. H., 55, 69
Bowler, R., 25
Boyle, D., 63
Breitenfeldt, M., 185
Brenner, P. W., 55, 75, 81

C

Capolupo, A., 31
Castro, J., 3

D

Danielson, J. R., 171, 199
Delahaye, P., 185
Dietrich, M. R., 25
Driscoll, C. F., 89
Dubin, D. H. E., 89, 121
Durand de Gevigney, B., 55, 63, 69, 81

E

Eades, J., 157
Eggleston, D. L., 102
Ennever, P., 63
Enomoto, Y., 163

F

Fletcher, R. S., 11

G

Gao, H., 3

H

Ha, B., 39
Hahn, M. S., 55, 63, 75, 81

Herfurth, F., 179
Herlert, A., 185
Himura, H., 47, 63
Hori, M., 157
Horváth, D., 157

I

Imao, H., 157
Isobe, M., 47

K

Kameyama, S., 19
Kanai, Y., 163
Kawai, Y., 108
Kester, O., 179
Killian, T. C., 3
Kiwamoto, Y., 108
Kluge, H.-J., 179
Komaki, K., 157
Kono, A., 19
Koszudowski, S., 179
Kotelnikov, I., 130, 136
Kurcz, A., 31
Kuroda, N., 157
Kurz, N., 25

L

Lefrancois, R. G., 75

M

Maero, G., 179
Marksteiner, Q. R., 55, 63, 75, 81
Marler, J. P., 39
Martin, X. S., 55
Masamune, S., 47
Mendez, J. M., 63
Mohri, A., 157, 163

N

Nagata, Y., 157
Nakamura, K., 47

O

O'Neil, T. M., 89, 114

P

Paroli, B., 96
Pedersen, T. S., 55, 63, 69, 75, 81
Petri, J., 191
Pozzoli, R., 96, 136

Q

Quint, W., 179

R

Rolston, S. L., 11
Romé, M., 96, 130, 136

S

Saitoh, H., 163
Sakawa, Y., 19
Salacka, J. S., 25
Sano, F., 47
Sanpei, A., 47
Schwarz, S., 179
Schweikhard, L., 185
Shibata, M., 157
Shoji, T., 19
Shu, G., 25
Shullman, M., 63
Smoniewski, J., 39
Stoneking, M. R., 39
Sturm, S., 185
Sugimoto, D., 47
Surko, C. M., 171, 199

T

Torii, H. A., 157

W

Weber, T. R., 171, 199
Wenander, F., 185

Y

Yamazaki, Y., 157, 163
Yeliseyev, Y. N., 142

Z

Zhang, X. L., 11